Subliminal Seduction
How the Mass Media Mesmerizes the Minds of the Masses

FIRST PRINTING

Billy Crone

Copyright © 2021
All Rights Reserved

Cover Design:
CHRIS TAYLOR

To Rhett Falkner.

*Thank you for being
one of the first people
to ever reach out to me
and pull me out of
the world of illusion
I was living in.*

*Unknowingly
I was once a slave to lies
But you showed me
the Way the Truth & the Life
Through Jesus Christ.
And now I am not only set free
but for all eternity.*

*Thank you for caring
and loving me enough
to tell me the truth.
For this I am
eternally grateful.
I love you.*

Contents

Preface vii

1. *The History of Subliminal Technology* 9
2. *The Methods of Subliminal Technology* 29
3. *The Manipulation of Newspapers* 105
4. *The Manipulation of Radio & Music* 131
5. *The Manipulation of Books & Education* 199
6. *The Manipulation of Television* 265
7. *The Manipulation of Social Media* 305
8. *The Response to Subliminal Technology* 377

    How to Receive Jesus Christ 395
    Notes 397

Preface

Before I became a Christian, I had a roommate that turned me on to a movie that came out in 1988, five years prior to my salvation in 1993, called, *They Live*. As a non-Christian, I can still remember how I was literally mesmerized by that movie premise of how Aliens had secretly invaded the planet and hijacked various media technologies to subliminally put the whole planet to sleep and cause them to obey whatever the Aliens wanted them to. Then, it was only the lead character who had stumbled across a set of special glasses that enabled him to see through this world of illusion created by the Aliens that gave him the ability to warn others and put a stop to it. I recall feeling, even back then, there has to be more truth to this than actual science fiction. Was it just a movie for our entertainment, or was there a profound message for all of us to receive? Now fast forward to several years after I was saved, sure enough, I began to stumble across shocking evidence that *They Live* is actually how *we live*! At first it sounded preposterous, but frankly, the more I went down this rabbit hole, the more evidence began to pile up and become undeniably overwhelming. I couldn't believe it, yet I dare not deny it. I didn't want to become a mindless slave like the rest of the planet let alone go back to sleep after what I had seen. I realized it was my duty to warn people that we really live in a world of illusion created for us by a handful of, not Aliens, but rich Elites, who are manipulating us for their own nefarious purposes. I then realized that even my fellow Christians were being affected, so I knew it was time to take action. Therefore, what you are about to read in the pages of this book is not only the cold hard facts about this grand global media matrix conspiracy that's been foisted upon us, but most importantly, you will be shown the way out to true and lasting freedom. One last piece of advice; when you are through reading this book, will you please READ YOUR BIBLE? I mean that in the nicest possible way. Enjoy, and I'm looking forward to seeing you someday!

Billy Crone
Las Vegas, Nevada
2021

Chapter One

The History of Subliminal Technology

Thank you for your courage to join us in our latest study entitled, "Subliminal Seduction: How the Mass Media Mesmerizes the Minds of the Masses." Right out of the gates, you can tell, with that eclectic title, that this comes across as yet another seemingly bizarre or outlandish topic for us to deal with. In fact, even more so than some of the other books we have done in the past. So, naturally, you might be wondering, "Why in the world are you talking about this topic at a time like this and how does this fit in with what's going on in our world today? Subliminal Seduction. Are you serious? Have you fallen off the deep end on this one?"

At the outset, I agree with you, it seems it is a bit farfetched. However, as you will see, and I think eventually even agree, once you look at the evidence, as wild as it appears, Subliminal Technology did not disappear in the 1950s like we were told. Rather, ever since then, this deceptive technology has actually been used on us, right on up to our present day, and it's being used on a massive scale, worldwide, to manipulate, mesmerize, and control the minds of the masses, for some pretty nefarious reasons by the way.

And for those of you who might not be familiar with Subliminal Technology, let alone its history, let's take a little stroll down memory lane, shall we?

Cheddar News: *"Every day the average American is exposed to between 4,000 to 10,000 ads. We are bombarded by them while we surf the net, watch TV or even while we are walking around. But what if there are more ads, ones that we don't consciously consume? What if advertisers found a way to tap into the subconscious? For example, as I was just talking, did you see that? It's just a frame long.*

(A cartoon of passengers on a bus, reading papers, but across the screen it has in large print SUBSCRIBE.)

Just short enough that most people would not even notice. This is an example of a visual subliminal message and did you just hear that? Let's bring nano vision to the front.

(Background noise is repeating, SUBSCRIBE.)

This was playing at intervals during the last minute. It is an example of an audio subliminal message.

On September 12, 1957, a market researcher held a press conference to present his findings of a resent experiment. The man was James Vicary. In his experiment, audiences were shown the film, 'Picnic'. Flash frames were littered throughout the movie, with phrases like 'Hungry? Eat Popcorn and drink Coca Cola'. Each frame lasted $1/3000^{th}$ of a second. Far too quick for a viewer to notice. Vicary claimed that the sales of Coca Cola and Popcorn had gone up 18.1% and 57.8% respectively.

Journalists began publishing articles claiming that America had entered the age of Big Brother. The dystopian world of 1984 had arrived. But there was a problem. Vicary revealed that the data was too small to be meaningful."[1]

Or was it? And this is where it gets interesting, because ever since James Vicary came back in 1962, five years after the initial experiment, where he said that the results of Subliminal Messaging upon people were quote, "Too small to be meaningful." Everybody, by and large, in the media and advertising industries, began to come out and say, right on up to this day, that this was all just a big hoax, that there's nothing to see here, there's nothing to be afraid of, we've done our own experiments on this technology and trust us, it's not true.

Okay. To me, this is like asking the fox to guard the henhouse! The very same people who could potentially use this Subliminal Technology on the masses for various nefarious purposes, are the very same people telling us to trust their own so-called independent studies on this technology and take their word for it, that there is nothing to worry about, they would never do this to us, and blindly take them at their word, go back to sleep, assume it all went away. I don't think so! And I would say that because one, Subliminal Technology, "can" and "does," influence people. It is a real technology, as you will see here in a second. And two, there are countries today that have actually banned the use of Subliminal Technology. So, my question is, "Why would you ban Subliminal Technology if there's nothing to worry about?" Let me show you what I mean.

SciPsych Show: *"Subliminal Perception is for sure a thing we definitely can react to a stimulus even when we can't consciously perceive it, which is different from superliminal perception. Things that we do consciously perceive, even if we don't pay direct attention to them, like product placement.*

The line in between those two is known as subjective threshold. Then below that is the objective threshold. The level at which we do not perceive or react to the thing.

Subliminal Perception Research dates back to a book published in 1898 when a psychologist was looking to confirm the idea of a so-called sub-waking self.

In a few experiments he asked a few dozen participants to read some numbers or letters on cards, but he held the cards so far away that they could only see a blur or small dot. When forced to choose, participants could usually distinguish between numbers and letters and they got it right around two-thirds of the time. They actually did better than chance at guessing exactly what was on each card, all of which suggested they were perceiving the images on some level. Even though they thought they were just guessing.

In a 1951 study in the Journal Psychological Review, found an even clearer evidence by conditioning people to associate certain nonsense words with an electric shock. Later when the words were shown to them too briefly to be consciously seen, the researchers measured greater electrodermal activity for words associated with the shock. That is a slight change in how well skin conducts electricity, which is associated with sweating. In other words, even though subjects believed they hadn't seen anything, their bodies still anticipated the jolt.

So, we know subliminal perception is real. If the Subliminal Advertising isn't technically illegal in the US, while Australia and the UK have laws against it, the US doesn't forbid advertisers or networks from using it."[2]

      Gee, I wonder why? And I also wonder why countries have put bans upon this kind of technology, if there was nothing to worry about and it's all just a big giant hoax, as they would have us believe. It doesn't make sense to me. Does it to you? Unless of course, it really does have an effect on us, as you just saw, and they really do use it, right on up to present day. In fact, let me give you a couple of easy, current, verifiable examples. Starting with the Subliminal Messaging that is embedded in company logos.

Business Insider: *"Subliminal Messages are everywhere, especially in corporate logos. A few examples of these are like the Tostitos logo which contains much more than their name, inside the logo are two people eating Tostitos.*

Inside the Hershey Kisses Brand, you can find a hidden Kiss between the K and the I.

When you look at the NBC logo between the pink and the purple wings of the peacock you can see where a hidden silhouette of the peacock lives.

The Milwaukee Brewers logo is much more than a baseball glove. The team's initials are hidden inside.

The smiley face on the Goodwill sign is actually forming a letter g in the name.

If you look at the Pittsburgh Zoo and PPG Aquarium sign you will see the silhouette of a tiger and an ape.

Formula One Racing has a secret buried in its logo. If you look at the negative space, you will see a large #1.

The FedEx logo hides an arrow in its negative space.

The Tour de France logo contains a well-integrated biker.

The Sun Microsystems logo is a perfect ambigram. It can be read from any angle. Also, the graphic doesn't actually include an S.

The Baskin Robbins logo cleverly uses the company's initials to advertise its number of ice cream flavors, the 31 is inside the BR initials.

The cleverness of this logo is twofold, Amazon. The arrow points from A to Z, referring to all that is available on Amazon.com and it doubles as a satisfied smile (with dimple).

The Toblerone logo is a tribute to the Swiss town where the chocolate was developed. Inside is a dancing bear in the mountain.

Some think the LG logo is a Pac-Man reference. The smiling, winking face is more apparent but only slightly.

The logo for Sony's computers represents the brand's integration of analog and digital technology. The 'VA' is designed as an analog waveform. The 'IO' is binary code."³

That's just company logos. What else is going on out there? Well, a lot, as you are going to see here in just a second. And that by the way is from Business Insider. This is not Joe Schmo dot com or Wacky Conspiracy Guy dot org or I Lost My Mind dot edu. It's really out there and they even admit it, whether you realize it or not. But again, that is just the tip of the iceberg with this Subliminal Messaging that's going on in our world today. It's not just being used in company logos, but even in company advertising. Look at these examples, again, from the Business Insider.

Business Insider: *"Subliminal sex messages in advertising: The classic Coca Cola image has two hidden, erotically situated women in it, located in the ice at the top of the can and a spot of ice just below the second line on the picture of the bottle on the can.*

A Palmolive ad with a subtle hint of something more. Take a closer look at the arm holding onto the woman's leg. Who does it belong to?

The Need for Speed II Xbox game cover has a secret. Take a look at the buildings in the distance. When you turn the cover sideways, the letters in the building lights spell SEX.

In 1990 Pepsi rolled out a line of 'cool cans' for the summer season. One of them got a lot of attention. When you stack them on top of each other, they also spell SEX in the lines that are drawn around the cans.

Kids and adults alike took notice when Skittles released their berry explosion flavor. Take a close look at the candy's new title. Berry

Explosion but with a red Skittle with an S on it in front of the word Explosion."[4]

Right there in plain sight, whether you realized it or not. Again, from the Business Insider, not from My Brains Fell Out on to the Floor and I Can't Find Them Anymore dot tv or something to that effect. And did you notice, they were using "immoral" means to get their Subliminal Messaging across for you and me to buy their products? And believe me folks, that is completely tame compared to what is really going on out there. I cannot, in good conscience, show you the other much more graphic sexually immoral ways they are doing it as well. Tons of examples are out there from reputable sources of Subliminal Messaging with sex acts, body parts, sexual innuendos of all kinds, just to name a few. Do your own research. It's sick!

In fact, speaking of sickness, they are doing this same immoral Subliminal Messaging not only in advertising, but even in kid's cartoons. Look at these examples from Disney.

Narrator: *"In case you didn't know, Disney is big on hidden images, references and even Subliminal Messages. Let's take a look at some of the best. The Lion King. At one point in the movie, Simba is having a rough time of it. As he releases a big sigh, he lays down on a cliff looking up at the night sky. As he does this a paw shoots out petals into the night breeze. Usually, this would be nothing but a nice visual, but these petals seem to spell out a word that should only be mentioned by adults. SEX.*

Known to many as the most beautiful cartoon character ever to have existed, Jessica Rabbit was introduced in 'Who Framed Roger Rabbit?' A seductive songstress that had quite a big part to play in the movie. But her legacy goes beyond that. She has inspired countless Halloween costumes, merchandise and even a line of makeup. There was nothing quite like Ms. Rabbit before. And there hasn't really been anything like her since. Grown men were suddenly developing crushes on an animation. The world truly went bonkers with Jessica and continues to do so. That is why the scene you are about to read about is one of the most paused moments in movie history. When the car sequence crashes, Jess is flung out of the car alongside some other characters. It seems quite normal when you play it at the normal speed, but if you slow it down you can clearly see that Jess's dress is going everywhere. Thankfully it's not detailed, but it is still a moment that could have been handled better. Maybe the animaters made a mistake thinking that no one would slow it down to take a peep in the years to come.

'Ratatouille'- the innuendo in this one is out of control. Ratatouille is a little more suitable for dirty jokes anyway, since the main characters are adults. But when Alfredo attempts to come clean about bringing a rodent into the kitchen, he struggles to get his words out. During the difficult exchange, he steps up the reveal to tell about his tiny little..., but before he can say the word rat, Collette drops the wickedest glance right at it. Somebody animated that glance and that facial expression.

'Rescuers Gone Wild'- not all the dirty bits were originally intended. In this 1999 VHS release of the 'Rescuer's, some hilarious or sometimes kind of sick, they managed to insert an image of a topless chick into the middle of a chase scene. Blink and you will miss it.

'The Lion King' – Mufasa's Face, not actually all the dirty stuff is in the movie, but the movie poster of Mufasa's face bears an outline that could, from a certain angle, look like a naked chick. Did they realize this when they made it? Was it a little joke inserted by the illustrator? Or could it just be the mind of a fresh minded internet user? We'll let you decide.

'Tangled' – All about the Hair, another poster and another Subliminal Message. It's a miracle we have reached into adulthood without our subconscious minds being completely corrupted by this endless train of filth. Who are we kidding? We are all totally disturbed. Now look, not everyone agrees with this one, but when you take a look at the Tangled poster, SEX, a big metaphor staring you right in the face. It is written in the hair wrapped around the fellow in the picture."[5]

Whether you, or your kids, realized it. In kid's cartoons! Why would you do that? That is sick folks! And that's just Disney! Who knows what else is going on out there in the cartoon industry with other companies! And believe it or not, there's even much more sexually graphic examples of Subliminal Messaging going on in kid's cartoons than what you just read. But again, in good conscience, I cannot, and frankly, refuse to show them to you. It is appalling.

But as you can see, this is really going on, I am not making this up, and it's not stopping with kids either. They not only use this graphic immoral adult Subliminal Messaging with kids, but they also do it with adults as well, on a massive scale! In fact, they use all kinds of Subliminal Techniques to manipulate audiences purposely, emotionally, and mentally for "their" desired effect, and most people have no clue. Let me give you just a couple of examples of how they do just that.

Narrator: *"One of the weirdest story-telling tricks employed by your favorite TV shows is Subliminal Messaging. The art of sneaking little clues, hints, or imagery on screen that a viewer may not notice the first, second or even the third time through, get ready to have your brain snapped around like a rubber band when you see how ingeniously they have been messing with your head. You may never trust TV again.*

'How I Met Your Mother' – As a three-camera sitcom, 'How I Met Your Mother' certainly didn't employ the satiric magic of Subliminal clues on a regular basis. But it did make clever use of them in an episode, 'Bad News.' If you were paying attention, you will notice in the first scene the number 50 is right there in the middle of the screen, kicking off the countdown that runs through the entire episode, from the label on a beer, (36) to a book cover, (25) to a sequential set of lottery numbers. (32, 31, 30, 29, 28, 27) If you are sharp eyed enough to notice the number on the jacket (10) reveals it is getting closer and closer until the countdown reaches (1). After building up that kind of subliminal dread about the unstoppable march of time, the news is revealed and as expected it is quite bad.

'Breaking Bad' – When future generations look back, 'Breaking Bad' may be regarded as the Shakespearean epic of our time. It had compelling character transgression, fantastic writing, great acting, and of course Subliminal Messaging. Every item, scene, and moment in 'Breaking Bad' is there for a reason, particularly when it comes to the show's color. The color code has been analyzed in major publications like Slate and Time, revealing that any color that a character would wear on the show would hint on what was going on in their head. When we first meet Walt, he is a mild-mannered creampuff that wears the muted colors of a potato sack. But when Walt takes the identity of Heisenberg and he starts getting drunk on power, his wardrobe changes to stronger colors only moving back to milder khaki when he has moments of insecurity.

'Fringe' – One of the creepiest feelings that you can experience is the sensation that you are being watched. And if you have ever had the urge to look over your shoulder while you are watching 'Fringe' there is a reason for that. Hidden in every single episode, yes, every episode through multiple seasons, an observer, a bald man in a business suit lingers in the background and watches whenever weird stuff is going on. If that wasn't creepy enough, Fox marketed the observers reveal on 'Fringe' by having an observer week, where observers popped up in all of Fox's programming. Appearing in places like 'American Idol' and even a NASCAR race and an NFL game with no explanation offered. It's a long way to go to creep out your audience.

'Cloverfield' – Did you know that 'Cloverfield' contains a 'King Kong' cameo? JJ Abram's monster movie is known for the shaky, stuttering, handheld camcorder and in between cuts and flickers the director inserted single frames from monster movies like 'King Kong' and 'The Beast from 20,000 Fathoms' and 'Them.' Whether they were put there to disturb the audience without them realizing it or as just a homage to the kind of movies that inspired 'Cloverfield', these splashes are a fun Subliminal secret that can only be seen if you slow the scene down.

'Irreversible' – Subliminal Messaging isn't always about advertising. Filmmakers use it to create a certain emotional affect in their audience *without them consciously knowing why they felt that way. The first 30 minutes of 'Irreversible' contains an extreme low frequency background noise that is virtually inaudible to humans but capable of causing uneasiness and nausea. He deliberately inserted it into his soundtrack to cause an insatiable feeling of disgust while watching his movie."*[6]

Total manipulation, whether you realized it or not. And notice, it wasn't just methods of subliminally inserting graphic sexually explicit imagery into these forms of media, but even inserting Subliminal placement, Subliminal colors, and even Subliminal sounds, that you don't even know are there, to manipulate people in a variety of ways! Did you know this was going on? Is there a disclaimer at the bottom of these media formats telling us what they are doing, so we have a heads-up knowledge in order to choose whether or not we want to subject ourselves to this obvious admitted manipulation? Why not? I thought you said there's nothing to worry about and you'd never use this kind of technology on us, there's nothing to worry about, and it's all just a big hoax! I don't think so. They even admit it, they are using it on us, right on up to this very day in a variety of ways!

And so that naturally brings us to the next question, "Does all this Subliminal Messaging really effect people?" Well, yes of course, because we already have proof of that in some of what we've already shown. In fact, in the last example they even admit that they use these Subliminal techniques, "to create the desired effect" upon their audiences. But let's take a look at some more proof that Subliminal Technology really does manipulate the minds of the masses.

Narrator: *"It has been estimated that across all media each day we are exposed to over 360 adverts and with the average American watching over 4½ hours of television a day the potential impact of Subliminal Messaging is huge. But what exactly counts as Subliminal? A common definition states that a stimulus is Subliminal when it is below the threshold of conscious awareness and cannot be subjectively identified from an alternative stimulus.*

For example, one study participant was shown pictures of a regular polygon for one millisecond, way too quick to be consciously aware of it. Afterwards, the participants were presented with pairs of figures, one that they had been exposed to previously and one which was new. They were asked to say which shapes they had seen before and which shape they preferred. Although they couldn't say which shape they had seen before, with guesses that varied no better than chance, they showed an increased preference for the shapes they had unconsciously perceived previously.

In another study, students were asked to write down three ideas for possible research projects. They were then exposed to the face of either a smiling or scowling professor. The students were not aware that they had been exposed to these faces and reported seeing only a flash of light. They were then asked to rate the quality of the research ideas they had proposed. Those who had been exposed to the scowling face rated their own ideas less favorably than those that had been exposed to the smiling face.

Perhaps the most interesting in the study of unconscious awareness, people have been shown to remember specific events while under general anesthesia. Patients were given headphones which played repetitions of a series of single words. Following surgery, they reported no memory of any particular words, however, if presented with word stems, such as GUI or PRO, they were much more likely to complete the words with those which were played to them whilst unconscious.

So, if the tapes included the words guide or proud, they were more likely to complete the word stems with these words compared with others. These effects however, only lasted up to 24 hours, with the best results from the tests taken immediately following the surgery. These experiments demonstrate the distinction between conscious and unconscious perception, with the latter leading to more automatic reactions. What about advertising? Real-world studies are hard to come by given the ethical implications."[7]

In other words, have fun trying to get these guys to admit they are really using this on us because of the obvious "moral" and "ethical" violations it presents to us. And yet, contrary to the narrative that is being fed to us, there are actually scientists out there, even besides the evidence you've seen so far, that fully admit Subliminal Technology really does have an effect on people. Here is just one example from the UK.

"Subliminal advertising really does work, claim scientists who found that people subconsciously respond to flashed messages – especially if they are negative.

Researchers found that briefly displaying words and images so quickly that people do not even consciously notice, does nevertheless change their thinking.

They found it was particularly effective with negative images and words which could alter a person's mood. British researchers have shown messages we are not aware of can leave a mark on the brain.

A team from University College London, funded by the Wellcome Trust, found that it was particularly good at instilling negative thoughts.

'There has been much speculation about whether people can process emotional information unconsciously, for example pictures, faces and words,' said Professor Nilli Lavie, who led the research.

'We have shown that people can perceive the emotional value of Subliminal Messages and have demonstrated conclusively that people are much more attuned to negative words.'

In the study, published in the journal Emotion, Professor Lavie and colleagues showed fifty participants a series of words on a computer screen.

Each word appeared on-screen for only a fraction of second – at times only a fiftieth of a second, much too fast for the participants to consciously read the word.

The words were either positive (e.g., cheerful, flower and peace), negative (e.g., agony, despair, and murder) or neutral (e.g., box, ear or kettle).

After each word, participants were asked to choose whether the word was neutral or 'emotional' (i.e., positive, or negative), and how confident they were of their decision.

The researchers found that the participants answered most accurately when responding to negative words – even when they believed they were merely guessing the answer.

Professor Lavie believes the research may have implications for the use of Subliminal Marketing to convey messages, both for advertising and public service announcements such as safety campaigns.

'Negative words may have more of a rapid impact,' she said. 'Kill your speed' should be more noticeable than 'Slow down.' More controversially, highlighting a competitor's negative qualities may work on a Subliminal level much more effectively than shouting about your own selling points.'

Subliminal Advertising is not permitted on TV in the UK, according to the broadcasting regulator Ofcom. However, there have been a number of cases where the rules have been stretched."[8]

In other words, they are doing it anyway. And yet, this is the constant challenge with exposing what is really going on with Subliminal Technology and how it is really being used on the masses to mesmerize their minds. As soon as anybody comes out with their own independent study on technology, demonstrating the clear real-world effects of Subliminal Messaging, like the one you just saw, the rest of the industry downplays it, pounces on it, denounces it as yet another hoax, or simply says it's nothing but the results of faulty research. Really? We already saw it is out there, still being used on us in a multitude of ways, they even admit it. This denial is the same ol' tired line, or should I say, lie, they've been using on us since the 1950s!

And then, on top of all that, for even more proof, they even have a whole field of study on this kind of manipulative Subliminal Technology called "Neuromarketing." It is defined as:

"A commercial marketing communication field that applies neuropsychology to market research, studying consumers' sensorimotor, cognitive, and affective response to marketing stimuli, using brain-scanning technology – such as MRIs and electroencephalography (EEG) – to observe how people's brains respond to a specific ad, packaging design, product design, etc. Marketers take the results of the scans and use them to create marketing consumers will find more appealing or motivating." [9]

In other words, they are using this science to get inside your brain to figure out how best to improve their techniques to manipulate you into buying their stuff! Now, wait a second! I'm supposed to believe you when you say Subliminal Technology is a hoax and there's nothing to worry about and you'd never use this on us when you have a whole field of scientific study out there dedicated to improving its results on us? I don't think so!

And here's my point. Remember, we are just getting started with this study. I believe the use of this obvious manipulative and mesmerizing Subliminal Technology, that really is being used on us, is not only clearly affecting people's minds, but it is also clearly affecting their behavior. In fact, the timing of this technology is spot on! Because if you read the Bible, it warned about a specific type of immoral and egregious behavior that would rapidly ramp up when you're living in the Last Days. Let me show you that passage.

2 Timothy 3:1-5 "But mark this: There will be terrible times in the last days. People will be lovers of themselves, lovers of money, boastful, proud, abusive, disobedient to their parents, ungrateful, unholy, without love, unforgiving, slanderous, without self-control, brutal, not lovers of the good, treacherous, rash, conceited, lovers of pleasure rather than lovers of God – having a form of godliness but denying its power. Have nothing to do with them."

So here we see, nearly 2,000 years ago, where the Bible clearly warns us that one of the major unfortunate characteristics of the Last Days society is that it's going to be a society filled with absolute unadulterated, ever increasing wicked behavior. It says people at that time are going to be selfish, greedy, boastful, prideful, abusive, disobedient, ungrateful, unholy, unloving, unforgiving, slanderous, out-of-control, brutal, evil, treacherous, rash, and conceited! And tell me that is not our world today! Every single one of them!

And so, my obvious question is, "How in the world did we get into this shape so fast as a society? I mean, it was not that long ago that we didn't have these kinds of behaviors in our world. Not on this scale. They weren't commonplace like they are now. So how did this happen?

In fact, before I answer that, for a reference point, it wasn't that long ago that the leading disciplinary problems in schools were, get this, talking, chewing gum, making noise, running in the hallways, getting out of place in line, wearing improper clothing, and not putting paper in the wastebaskets! Oh, those rebels! In fact, not that long ago, the biggest trouble people got into was like what happened here with Beaver in *Leave it to Beaver*. Remember this?

A clip from Leave it to Beaver: *The clip opens with Beaver's mom asking him, "How was the movie?"*

The Beav has a guilty look on his face as he answers, "I didn't go to the movie."

His dad is rather concerned at his answer and asks, "You didn't go to the movie?"

Now, Beaver knows he is in big trouble and answers his dad, "No sir, I went yesterday when I wasn't supposed to."

The dad replies, "Oh, is that so."

Beaver then says, "Yes sir, and I won a racing bike with a genuine leather seat, and I hid it at Larry's, and I was going to make believe that I won it today, but I couldn't, so that is why I am telling you what happened."

"Well," the dad says, "When did you decide to tell us about it?"

Beaver answers, "When I was walking the bike home from Larry's."

At that point Wally chimes in, "Yeah dad, it's too big for him to ride." The audience laughs.

"Well, Beaver, I'm glad you decided to tell us the truth." His dad replies. "Of course, you know you can't keep the bike you won when you were being disobedient. We will have to find something to do with the bike."

But Beaver seems to have the problem solved when he says, "Larry and I have already found something to do with it. I walked it back down to Larry's house and then Larry and I took it down to a church."

His dad asks, "To a church?" Beaver explains, "Larry saw what a church does with babies."

Out of curiosity the dad asks, "Does what?" Beaver tells him that they left the bike on the steps of the church with a note. "We just hope that someone nice adopts it."

At that point the dad tells him, "I am glad you realized you couldn't keep the bicycle, but there is still the matter of you being disobedient, isn't it?"

Beaver answers, "Yes sir."

So, his dad tells him that he is going to have to stay away from the movies for a while, for 2 weeks.

Beaver answers, "Yes sir" and turns to leave the room.[10]

What a rebel! Can you imagine the audacity of Beaver there? What wickedness! What horrible rotten behavior was demonstrated by him! Our society's falling apart! Yeah, I'm joking. How we wish it were only that bad today. Look around. Now we have to deal with these top problems at school today, drug abuse, alcohol abuse, pregnancy, suicide, murder, guns, robbery, rape and metal detectors everywhere.

So again, my question is, "How in the world did these behaviors change so fast?" Well, believe it or not, I believe the media, including the use of Subliminal Technology, has a very large part to play in it.

Now, I will say this, as a disclaimer of sorts, people are still responsible for their own actions and behaviors. They cannot and should not go around blaming others for what they do, including the use of Subliminal Technology on them. No, they need to own up to it.

But I do believe that this type of Manipulative Technology is "influencing" people to "make that choice" whether they realize it or not, to engage in this type of immoral egregious Last Days behavior. And again, that is why we entitled this documentary, *"Subliminal Seduction How the Mass Media Mesmerizes the Minds of the Masses."*

In fact, I also find it interesting that *Leave it to Beaver* with its mild seemingly innocuous behavior as compared to today, came out in 1957, the exact same year James Vicary came out with his Subliminal Movie Popcorn Coke experiment. And now, over 60 years later, ask yourself, "What would happen to a society that has been theoretically subjected to a constant ever-increasing insertion of immoral Subliminal Messages into their advertisements, movies, television, and even kid's cartoons?" I think it would help explain why there is such an increase of immoral behavior in people in such a very short period of time, just like the Bible warned would happen in the Last Days. Again, the timing is spot on. It is not by chance.

In fact, to prove my point even further, let's now move on to the Methods of Subliminal Technology. What you are about to see is going make it even more clear as to why our society is going down the tubes like the Bible predicted.

Chapter Two

The Methods of Subliminal Technology

"A few months before I was born, my dad met a stranger who was new to our small town in Tennessee. And from the beginning, Dad was fascinated with this enchanting newcomer and actually invited him to live with our family. The stranger was quickly accepted and was around to welcome me into the world just a few months later.

As I grew up, I never questioned his place in our family. In my young mind, each member had a special place. My brother, Bill, five years my senior, was my example. Fran, my sister, gave me an opportunity to play 'big brother' and develop the art of teasing. My parents were complementary instructors—Mom taught me to love the Word of God, and Dad taught me to obey it.

But the stranger was our storyteller. He could weave the most fascinating tales, adventures, mysteries and even comedies. He could hold our whole family spellbound for hours every single evening. In fact, if I wanted to know about politics, history or science, he knew it all. He knew about the past, understood the present, and seemingly could even predict the future.

And the pictures he could draw were so lifelike that I would often laugh or cry as I watched. He was like a friend to the whole family. He took Dad, Bill, and me to our first major league baseball game. He was always encouraging us to see the movies and he even made arrangements to introduce us to several movie stars. My brother and I were deeply impressed by John Wayne in particular.

And the whole time, the stranger never seemed to stop talking. Dad didn't seem to mind but sometimes Mom would quietly get up while the rest of us were enthralled with one of his stories of faraway places, go to her room, read her Bible and pray. I wonder now if she ever prayed that the stranger would leave.

You see, my Dad ruled our household with certain moral convictions. But this stranger never felt obligated to honor them. Profanity, for example, was not allowed in our house—not from us, from our friends or adults. Our longtime visitor, however, used occasional four-letter words that burned my ears and made Dad squirm.

And to my knowledge the stranger was never confronted. My Dad didn't permit alcohol in his home, not even for cooking. But the stranger felt like we needed exposure and enlightened us to other ways of life. He offered us beer and other alcoholic beverages often. In fact, he made cigarettes look tasty, cigars appear manly, and pipes were very distinguished.

He talked freely, probably much too freely, about sex. His comments were sometimes blatant, sometimes suggestive, and generally embarrassing. I know that my early concepts of the man-woman relationship were influenced by the stranger.

As I look back, I believe it was the grace of God that the stranger did not influence us more. Time after time he opposed the values of my parents. Yet he was seldom rebuked and never asked to leave.

More than thirty years have passed since the stranger moved in with our family in our small town in Tennessee. He is not nearly so intriguing to my Dad as he was in those early years.

But if I were to walk in my parent's den today, you would still see him sitting over in a corner, waiting for someone to listen to him talk and watch him draw his pictures. His name? We always just called him 'TV.'"[1]

Now, that story not only has a surprise ending, but a very convicting one, doesn't it? And if we are honest with ourselves, most of us would admit that watching TV endlessly, hour after hour, probably isn't the best thing for our minds, is it? In fact, it reminds me of the saying, "How much TV would you watch if you had to chop wood to keep it going?" Think about it. Most modern media is a mindless time waster. In fact, it not only wastes our time, but it also puts waste into our mind! Which apparently is why Groucho Marx stated, "I find TV very educating. Every time somebody turns it on, I go into another room and read a book!"

But here's my point. Like Groucho Marx, how many of us will admit that ingesting countless hours of various forms of media is not just a time waster, it really is harmful to our lives? It is, "Junk in equals junk out." And not only that. But how many of us will actually do something about it?

I say that because right now, whether you realize it or not, you are like one of those people in that story. You are either like the Father who is in denial over the harm the stranger is causing, or like that son who listens to the stranger because a parent allows them to do it, or you're like that mother who knows it's wrong, but you still don't do anything about it!

And again, this is one of the many reasons why we are writing this book entitled, *"Subliminal Seduction: How the Mass Media Mesmerizes the Minds of the Masses."* Our desire is that hopefully, after reading this, you will have the courage to change your viewing habits for the better, for your benefit. In fact, we will be ending on that point.

But for now, let us continue to explore where we left off last time with the question we posed, "Could it be that the media, including the usage of Subliminal Technology embedded throughout the media, really is responsible for the influence of immoral behavior that the Bible said would appear on the scene when you are living in the Last Days?"

Now again, as a disclaimer, people who are responsible for their own behavior and choices should not go around blaming others, including technology for their actions. You need to own up to it.

But my question is, "Does this Manipulative Technology, including Subliminal Technology, really 'influence' people to 'make' those 'choices' to engage in the types of immoral behavior the Bible talked about?" I think the answer is a resounding yes, especially when you are aware of the multitude of ways they are using this Seductive Technology on us, whether we realize it or not. So, let's dive into those methods, shall we?

The **1st method** Subliminal Technology is being used on us is **By Our Sense of Smell**. Let's take a "whiff" of what they're up to with this one.

You see, it is actually reported that Disney uses this Subliminal Smell technique in their theme parks by putting vanilla on cookie sheets and then heating them up in an oven. Why? Because people then start to "smell" the smell of "cookies baking" and come into the cookie shop for fresh cookies. But those cookies were baked somewhere else, but the "smell" of vanilla persuaded them to come in and buy the not-so-fresh "fresh-smelling" cookies! All a process of manipulation through our sense of smell.

Then, there is the not so obvious smells called pheromones. Researchers at the University of Chicago have demonstrated the existence of human pheromones or compounds that are produced by a person that can influence the biology or behavior of another. The new results have even been reported in the journal Nature, and it shows that people are indeed capable of being "led by the nose." *"This is a very exciting study that is going to make a lot of researchers sit up and take notice,"* said Dr. Charles Wysocki of the Monell Chemical Senses Center in Philadelphia. Other researchers predicted that, *"The discovery would throw wide open the door on pheromone research,"* and they said that *"Subliminal chemical cues very likely underly many human behaviors."*

In other words, they influence them, as this article also admits!

"Shoppers on the prowl for digital playthings unwittingly stumbled into a covert operation on the olfactory frontier. Riding up the escalator to the third floor of the Shops at Columbus Circle in New York, they encountered a scent.

It was not emanating from one of the many European tourists cruising the mall or escaping from a promotional event at a nearby store. No. It was the seductive smell of consumer electronics.

Samsung was conducting a test of its new signature fragrance in its Samsung Experience Concept Store. Researchers asked the shoppers leaving the store if they thought the scent was "stylish," "innovative," "cool," "passionate," or "cold," and, more important, whether the scent made them feel like hanging around the shop a little longer.

Nicole Snoeker, 25, was charmed. "I thought the store had just opened," she said. "It smelled very fresh and new." Plus, she even admitted, she probably lingered a bit more than she had intended. "I felt relaxed. It put me in the right mood. That's important in shops nowadays."

And the rival Sony Style Store has staged a preemptive strike in the odoriferous battle, with a shop scented with notes of mandarin orange and vanilla. "We wanted to add one extra dimension to differentiate our store from the rest," says Christine Belich, noting that the company is particularly interested in attracting female shoppers.

The company is also exploring a way to make the store's windows radiate the scent, so passersby might be lured inside to sniff out a laptop or new digital camera.

While vision is unquestionably our most powerful sense, when it comes to garnering an emotional response, scent is a much more powerful trigger. "Seventy-five percent of the emotions we generate on a daily basis are affected by smell."

And the average human being is able to recognize approximately 10,000 different odors. What's more, people can recall smells with 65% accuracy after a year.

And because of this, Westin Hotels recently began a test of White Tea, its new signature fragrance, in eight cities around the world, deploying the scent in the hotel's public spaces via a diffusing machine.

So, does it work? You bet it does! Dr. Eric Spangenberg, Dean of the college of business and economics at Washington State University, ran a test in a clothing store in the Pacific Northwest to determine how scent affected customers by gender.

He diffused a subtle smell of vanilla in the Women's Department and rose maroc (a spicy, honeylike fragrance that had tested well with guys) in the Men's. The results were astonishing. When he examined the cash-register tapes, he found that receipts almost doubled on the days when scent was used."

In other words, these Subliminal Smell techniques, attacking our senses, really do influence people's behavior! And just when you thought you've smelled it all, believe it or not, researchers are now working on ways to make your TV smell, no pun intended!

"Wake up and smell the TV. Virtual reality television to be a commercial reality in 15 years. Imagine watching a football game on a TV that not

only shows the players in three dimensions but also lets you experience the smells of the stadium and maybe even pat a goal scorer on the back.

The targeted "virtual reality" television would allow people to view high-definition images in 3D from any angle, in addition to being able to touch and smell the objects being projected upwards from a screen parallel to the floor.

"Can you imagine hovering over your TV to watch Japan versus Brazil in the finals of the World Cup as if you are really there?" asked Yoshiaki Takeuchi.

While companies, universities and research institutes around the world have made some progress on reproducing 3D images suitable for TV, developing the technologies to create the sensations of touch and smell could prove the most challenging, Takeuchi said.

Researchers are looking into ultrasound, electric stimulation and wind pressure as potential technologies for touch.

Such a TV would have a wide range of potential uses. It could be used in home-shopping programs, allowing viewers to "feel" a handbag before placing their order, or in the medical industry, enabling doctors to view or even perform simulated surgery on 3D images of someone's heart.

The future TV is part of a larger national project under which Japan aims to promote."[2]

In other words, they are going to be using it on us, and who knows what else. But as you can see, something doesn't smell right with this Subliminal Method. In fact, I'd say it downright stinks! But it doesn't stop there.

The **2nd method** Subliminal Technology is being used on us is **By Our Sense of Savor**. Let's get a "taste" of what they're up to with this one.

You see, when it comes to our sense of savor or taste, most of us may have already heard how cigarette manufacturers were caught red handed adding chemicals in cigarettes to make them more addictive, i.e., influence people's behavior through "taste" whether they realized it or not. In fact, the European Union allows over 600 additives to be used in the manufacture of tobacco products.

But what most people don't realize is that this Subliminal "taste" Method is not only used on those who like to blow smoke up in the sky, it's actually being done regularly to our food supply. In fact, many researchers admit that hardly any of us have ever known a true natural hunger pang. Rather, the food that we eat is laced with chemicals that cause us to want to eat more and more of the bad food within hours. For instance, it is now a well-known fact that MSG or Monosodium Glutamate is added to nearly all processed and manufactured foods as a flavor enhancer. The only problem is, it has no taste or nutritional value in and of itself. Free glutamate works by stimulating the taste buds and exciting neurons in the brain, creating the "illusion" of tastier food while suppressing bitterness in the mouth. And MSG is actually a drug and is very addictive according to toxicologist, Dr. George Schwartz. *"It intensifies pleasurable tastes and causes one to crave foods high in glutamate."* In fact, the more processed foods you eat, the more you crave them. Why? Because they disrupt the brain's natural chemical balance, and your natural hunger becomes distorted, causing you to now "crave" food.

And it's known that Aspartame or NutraSweet has some similar sour side effects as well. Believe it or not, Aspartame/NutraSweet is not very sweet in and of itself. It's actually a brain drug that stimulates your brain, so you "think" that the food you're eating tastes sweet. And it's used in all kinds of so-called diet products including sodas. However, an interesting side effect is that Aspartame/NutraSweet actually "causes" you to crave carbohydrates. So, so much for being "diet pop." Maybe we should call it "diet plump"! Total manipulation by our "taste!"

I know this is a hard truth to swallow, and it tastes downright rotten, but we are just getting started with these Subliminal manipulative methods attacking our senses.[3]

The **3rd method S**ubliminal Technology is being used on us is **By Our Sense of Sound**. Let's "lend an ear" to what they're up to on this one.

You see, another popular way to subliminally persuade someone is by using sound. In fact, if you recall, we saw this earlier in the study featuring how sound is being used subliminally in movies to make people purposely feel uncomfortable. However, this can be done in a variety of ways.

The first way you can subliminally manipulate someone with sounds is through a process that's called "back-masking." And this is when they make a backwards recording in music, on purpose, that's often used by musicians to subliminally influence their audience. And here's what they do. Vocals or other noises are simply recorded as usual and then flipped around backwards by a computer to be embedded "unintelligibly" into the music. However, it's still very intelligible to your subconscious mind and you will pick up on it.

In fact, it's so common place, this method of subliminal sound technology, that its already been used by bands like Pink Floyd, who actually made a parody of it. Their song "Empty Spaces" played in reverse says, "Congratulations. You've just discovered the secret message. Please send your answer to Old Pink, care of the funny farm, Chalfont."

In fact, even some so-called Christian bands are getting in on this parody of subliminal music. In Petra's song "Judas Kiss" played in reverse says, "What are you looking for the devil for, when you ought to be looking for the Lord?"

But unfortunately, not everyone uses this subliminal messaging in music for laughs. In the famous song from Queen, "Another One Bites the

Dust" some would say if you play it backwards it says, "It's fun to smoke marijuana" over and over again.

Oh, but rock artists aren't the only one's getting into this Subliminal influence using sound, so are pop artists like Britney Spears song "Hit Me Baby One More Time" in reverse seems to be saying, "Sleep with me. I'm not too young."

And last but not least, some artists use this Subliminal sound technique in their music for satanic purposes like the band Grim Reaper in their song, "Final Scream." In reverse it clearly and boldly says, "See you in Hell!"

But you might be thinking, "Hey, this is all just one big musical coincidence. There's no way this really has an effect on us." Well, if it is really all just "in our head" with no effect on us, then why have there been lawsuits brought against some of these musicians who use this Subliminal sound technology?

Furthermore, if it's really just "all in your head" with nothing to worry about, then why do storeowners use subliminal sound techniques to curb shoplifting in their music? Now, the technique they use is not "backmasking" or playing a message in reverse like we just heard with those musicians, but one that plays a message "below" (subliminal) or even "above" (supraliminal) our normal range of hearing. Yet, it nonetheless still is heard by our brains.

For instance, over 20 years ago, numerous department stores in the United States and Canada, installed what was called "the little black box," which mixed music and anti-theft messages. The quick repetition of "I am honest. I will not steal" 9,000 times an hour at a barely audible volume was able to curb shoplifting at one department store by 37% during a nine-month trial.

And retailers around the country are experimenting with "no shoplifting" audiotapes that endlessly repeat on a subconscious level using

Muzak, the directives, "Do not steal." One New Orleans supermarket reported that theft rates fell two-thirds from $50,000 yearly to just $13,000.

In fact, this Subliminal Method is not only used to curb theft, but even other "bad behavior" as well. In a Kansas City medical center, hidden audio messages are constantly piped through the sound system causing dramatic results. Smoking in the staff lounge is down 50% and angry patient outbursts in the crowded patient room are down 60%.

"According to the research conducted by Adrian C. North, a psychologist at the University of Leicester, "Unobtrusive music selected by store managers, business managers, and companies like Muzak can affect a person's thoughts and action without the person even knowing."

In fact, speaking of Muzak, let's go a little deeper into the purpose of the so-called "music service." Is it really just to pipe in music for businesses in an easy manner for the pleasure of their customers? Or is it there to manipulate them with sound?

"People listen to music for various reasons. Some people use music in order to increase relaxation. Others use music as a form of energy. Music is heard in cars, in homes, at shopping malls, and at dentists' offices, among many other places around the world.

Sometimes, a song gets into your head and you find yourself humming the tune all day long and then you realize that a stranger who had passed you hours ago had been whistling that song, or that you had heard 2 seconds of that song on your radio alarm that morning before pressing the snooze button.

This is the idea behind Muzak. In 1922, General George Squier invented Muzak, a type of music to deliver from phonograph records to workplaces via electrical wires. He realized that the transmission of music at the workplace increased productivity of his employees.

Soon after, there was a study that showed that people work harder when they listen to specific kinds of music. As a result, the BBC began to broadcast music in factories during World War II in order to awaken fatigued workers.

Muzak's patented 'Stimulus Progression' which consists of quarter-hour groupings of songs is the foundation of its success. Stimulus Progression incorporates the idea that intensity affects productivity.

Each song receives a stimulus value between 1 and 6 - 1 is slower and 6 is upbeat and invigorating. A contemporary, instrumental song full of strings, brass, and percussion (27 instruments in total) would most likely receive a stimulus value of 5.

During a quarter hour, about six songs of varying stimuli values are played followed by a 15-minute period of silence. A 24-hour plan is engineered to provide more stimulating tunes when people are the most lethargic - at 11 a.m. and 3 p.m. and slower songs after lunch and towards the end of the day.

Careful programming of Muzak has been proven to increase morale and productivity at workplaces, increase sales at supermarkets and even dissuade potential shoplifting at department stores.

More recently, a psychologist measured the influence of music on decision-making. He and his colleagues tested the effect of in-store music on wine selections at a supermarket by setting up a wine shelf with French and German wines. On alternating days, French accordion music or German pieces done by a Bierkeller brass band were played over a two-week period.

The results indicated that music did indeed influence shoppers' wine selections. When French music played, 40 bottles of French wine and 8 bottles of German wine were purchased. When German music played, 22 bottles of German wine and 12 bottles of French wine were purchased.

Researchers Charles Areni and David Kim have established a preference-for-prototypes model, which suggests that the mind is composed of closely packed, interconnected cognitive units which relate music and other structures and ideas. According to their model, music can stimulate the mind into thinking about ideas similar to the music. For example, French music conjures up images of France.

In addition, the speed of music can influence behavior. For example, several studies have been conducted which illustrate how fast music makes supermarket shoppers move around more quickly. Likewise, fast music causes diners to eat faster and slow music slows eating down (and leads to more drinks being purchased at the bar).

What is interesting about background music is that it is intended to be just that - noiseless noise. The concept of barely audible tunes affecting one's behavior leads to the question as to whether one's behavior can be manipulated by another individual without the person being aware of the manipulation.

According to research, stimuli below the threshold of conscious can influence thoughts, feelings, and actions without the I-function becoming involved or even knowing about it - that there is unconscious perception. It is likely that a number of other things can cause the same result.

As a result, people are not consciously recognizing that they are being manipulated by music when it is occurring."[4]

    Well, there you have it. Secular research admits that there is no way to know in totality all the ways we are being manipulated, subliminally when we are hearing or watching the "sound" of TV or listening to a radio or any other source of "sound" or music when we are at home or on public. Total manipulation! Somebody's trying to stick something between our ears, and it isn't' a box of Q-Tips either! But these Subliminal Sensory Methods do not stop there.

The **4th way** Subliminal Technology is being used on us is **By Our Sense of Sight**.

Speaking of Television, this is yet another popular method we are being Subliminally manipulated as we dealt with in the first chapter of our book in The History of Subliminal Technology. However, TV is not the only way they are manipulating us by "sight." Believe it or not, we haven't seen anything yet when it comes to this visual subversion. Let's see just how deep they'll go into our pockets.[5]

The **1st way** they visually subliminally seduce us is by **Stealth Marketing**.

Whether you realize it or not, Stealth Ads are simply commercials that fly in under the radar, tucked away in places that shouldn't have any commercials at all. And what is so seductive about it, is that they come in a variety of disguises.

The **1st disguise** is by **Commercial Branding**.

Commercial Branding is a "visual" Subliminal Technique used to seduce you into buying things, not with just a visual commercial in between the shows like we're used to, but rather, with a Subliminal visual commercial embedded "in" the show. They call these manipulative techniques, "Commercial Branding" or "Product Integration." And it's a huge arsenal of weapons that the modern media and advertisers rely upon to deliver Subliminal messages in order to get people to buy their products. This includes product placements in TV, news, movies, and talk shows, all with celebrities who are actually purposely paid endorsers. The whole process is carefully scripted, whether you realize it or not.

For instance, it wasn't by chance that an Apple computer was used by Carrie Bradshaw in *Sex and the City*, or the Ray Bans were worn by Tom Cruise in *Top Gun*, or the "Shaguar" Jaguar driven by Austin Powers in *Goldmember* and even Dougie the Pizza Boy spending weeks to try to deliver Pizza Hut pizzas to the *Big Brother* housemates and on and on it goes, just to name a few examples of this Subliminal Technique. In fact, they have even made whole shows dedicated to this Subliminal Technique of "hiding" commercials embedded throughout a program, like the so-called reality show, "Extreme Makeover Home Edition." In reality, it's all just one big giant ongoing weekly commercial for Sears, Furniture & Appliances Now, Pella Windows & Doors, Owens Corning, Lumber Liquidators, Wal-Mart, and Disney World Resorts, just to name a few!

And they not only do this with Television, but they also do it in movies as well. Let me give you just a few examples.

Watchmojo.com reports: *"We may not like being sold to but sometimes it works."*

Example - Clip from 'Pulp Fiction' – John Travolta and his partner, Samuel L. Jackson are riding in their car.

Samuel L. Jackson: "What do they call a Big Mac?"

John Travolta: "A Big Mac is a Big Mac, but they call it Love Big Mac."

Narrator: *"Welcome to Watchmojo.com. Today we are counting down the top 10 picks of product placement in movies.*

As an example - 'Back to the Future Part II' (1989) – The actress is tearing the seal off a frozen pizza package from Pizza Hut. 'Ummm, pizza.'

For this list we are looking at effective marketing strategies in movies that made product placement oblivious or even enjoyable. While shameless product placement has tendencies to annoy audiences, delicate and intelligent placement can get us up out of our seats and into our wallets. But either way, be sure to check out our list of the top ten shameless product placements in movies as well.

10). 'The Matrix' (1999) – Nokia. The man is opening an envelope at his computer with the word Nokia in large letters. Inside the envelope is a cell phone made by Nokia. As he is holding the phone, it rings. At the other end of the line, the phone has been dropped off the ledge of the multi-level building. A passerby picks it up and puts it in the trash, unbroken.

It would make sense that the makers of the ever-lasting Brix cell phone would have had a feature in this sci-fi classic. For the film, the Nokia 8120 was featured as the phone used to bounce in and out of the artificially generated reality. It was a snazzy phone with some appealing features at the time and best of all it made you look like you were taking a really important phone call.

9.) 'A Christmas Story' – Red Ryder BB Gun. If the threat of losing an eyeball wasn't enough to dissuade little Ralphy Parker from lusting after a Red Ryder Carbine Action 200 Range Shot model air rifle for Christmas, then it certainly wasn't going to stop kids in the audience. Even if you weren't a kid who liked playing with guns or even liked shooting things, by the end of the film you kind of wanted an air rifle made by Red Ryder.

Ralphy Parker: 'I think everybody should have a Red Ryder BB Gun.'

You probably didn't care about taking the chance of shooting your eye out either.

8). 'Transformers' – Chevrolet Camaro.

Main character: 'It's got racing stripes.' He is looking over the car while talking to the salesman.

Salesman: 'Yea.'

Although General Motors is all up and throughout the film and making what we call an ungraceful appearance the true stand out product placement is the Camaro. Bumble Bee, the autobot endears himself to human Sam Witwicky and then transforms himself from an old Camaro to a slick sexy concept Camaro. Making himself a hero like Shia LaBeouf.

Bad guy: 'You want to put the fate of the world on the kids Camaro?' That's cool.'

After being featured in the film, sales of the sporty yellow car skyrocketed. Making the Camaro the leader in the cool car department.

Optimus Prime: 'We live among these people now, hiding in plain sight, but watching over them in secret.'

7). 'Harold and Kumar Go to White Castle' (2004) – White Castle Burgers.

One of the characters: 'Just thinking about those tender little White Castle burgers, with those itty bitty grilled onions that explode in your mouth like flavor crystals every time you bite into one.'

The title of this slider franchise is in the name of the movie. Which seems like a pretty shameless plug to us. Two stoned high achievers set out on a mission to solve their White Castle cravings. Their mission seems to run parallel with many real-life munchies run. And the burger joint saw a large boost in sales following the release of the film.

Character: 'I want 30 sliders, 5 french fries and 4 large cherry cokes.'

The second character says, 'I want the same thing, except make it diet cokes.'

Never underestimate the buying power of college kids.

Character: 'That was the best meal of my life.'

6). 'The Italian Job' (1969/2003) – Mini Cooper.

*Character: 'You have a bunch of cars in there, right?' Mini Coopers?'
The salesman says, 'No.' The main character tells him, 'I'll give you 5-grand to put me into that car.'*

Audiences may not be really aware of these adorable European cars prior to the film's release. Although Mini Coopers were filmed in the 1969 original, most American movie goers weren't exactly familiar with the popularity overseas. That is until the 2003 remake. These retro inspired BMW manufactured vehicles were integrated into the plot making them stars in their own right. And having them seem so darn adorable in the chase scenes didn't hurt sales either.

5). 'Cast Away' (2000) – Wilson Sporting Goods.

Tom Hanks is on an island after being in a plane crash, and he seems to be the only survivor. He is talking to a ball that had washed on shore. That ball was the only thing that kept him from losing it being all alone.

Tom Hanks: 'Well, regardless, I would rather take my chance out there on the ocean than to stay here and die on this island for the rest of my life talking to a volleyball!!'

It is a story of a FedEx worker who survives a plane crash and ends up on a deserted island and believe it or not it was actually positive PR for the courier service. But Wilson Sporting Goods made out even better when Tom Hank's character fashioned a best friend out of a volleyball and a bloody handprint. He ends up humanizing an inanimate object. He also ended up advertising a sporting goods company for a solid portion of the film without it being annoying.

4). 'James Bond' franchise (1962) – Omega SA/BMW/Aston Martin.

*Instructor: 'Now, if you will see here, if you lift of the lid here, (on the gearshift) you will find a little red button. Whatever you do don't touch it.'
James Bond: 'And why not?'*

Being a debonair secret agent man apparently comes with some perks. Clearly you always get the girl, but you are also always decked out from head to toe with motor vehicles, in the finest of luxuries.

Instructor: 'Aston Martin calls it the Vanquish, we call it the Vanish.' He pushes a button, and the car that was invisible is now solid, right in front of him. And it comes with guns.

What advertiser wouldn't want to get in on that. When you are a luxury brand aligning yourself up with sophistication and it doesn't get more sophisticated than James Bond. Just look at his watch and his wonderfully exciting cars, do you need more proof?

3). 'Top Gun' (1986) – Ray-Ban/U.S. Navy. Nothing screams, 'I want to be a fighter pilot,' more than a bomber jacket and a pair of Ray-Ban aviator glasses. Especially since they were widely included in this Tom Cruise classic. The actor had already boosted sales for the shade maker years earlier in 'Risky Business.' But 'Top Gun' prompted a sales boost of 40% in just 7 months. However, no institution saw a bigger marketing impact than the U.S. Navy, which saw a 500% uptick in recruitment. Everyone wanted to learn the 'Top Gun' hi-five and take the highway to the danger zone.

2). 'E.T. the Extra-Terrestrial' (1982) – Reese's Pieces. Chocolate covered candies, and aliens do not necessarily go hand in hand. But you have to use something to lure an extra-terrestrial. Apparently, the role of the sweet bribe was supposed to go to M&M's who declined for whatever reason, leaving Hershey's new product Reese's Pieces to pick up the win. Reports say the peanut butter-based treat saw an 85% increase in sales in the wake of the film's success. So, we guess not only aliens like bite size candy.

Before we plug our top pick, here are a few of our honorable mentions.

'Lost in Translation' (2003) – Suntory Whisky. 'For relaxing times, make it Suntory time.'

'Gone in 60 Seconds' (2000) – Shelby Mustang GT 500. Nicholas Cage is driving one in this movie.

'Up in the Air' (2009) – American Airlines. 'I wasn't sure this even existed. This is the American Airlines concierge key.'

'Forrest Gump' (1994) – Dr. Pepper. Tom Hanks says, 'I must have drank me about 15 Dr. Peppers.' Then he belches.

'Dirty Harry' (1971-88) - .44 Magnum. Clint Eastwood says, 'This is a .44 Magnum, the most powerful gun in the world. It would blow your head clean off. You better ask yourself one question, do I feel lucky? Well, do you punk?'

1). 'Back to the Future' franchise (1985-90) – DeLorean Motor Company. This film franchise was notable for a number of reasons, creativity, the fun adventure, and the interesting plot. But the one thing that stood out for the audience was the futuristic vehicle at the center of the time traveling story.

Michael J. Fox: 'You made a time machine, from a DeLorean?!'

Doc: 'The way I see it, if you are going to build a time machine from a car, why not do it with style.'

With super cool gull wing doors, a silver chassis and an angular body style, it wasn't your run of the mill coupe.

Doc: 'When this baby reaches 88 miles per hour you are going to see some serious stuff.'

And even to this day, this car's exclusivity makes it one of the most sought after in the world."[6]

Don't worry. This stuff has no effect on you. It's all a hoax. Besides, they would never use it on you! Yeah, right! But that's not all. In one Hollywood movie alone, there were Subliminal placed ads for Target,

Philips Electronics, Starbucks, Motorola, Rolling Stone Magazine, Evian Bottled Water, McDonalds, Street Wear, Revlon, Kodak, Sega, and Ford. And why do they do this? Because it works folks, and they admit it!

    This is why advertisers pay millions of dollars to TV and movie producers to do this Subliminal product placement which in turn saves the movie producers tons of money on production costs. What's happening is that TV shows and box office movies are actually getting turned into one giant string of Subliminal Commercials to manipulate you and I into buying their products! Gee, I wonder why people are getting so greedy and materialistic these days! I wonder if it's influencing their behavior.

    But that's not all. Even the nightly news is getting in on this Subliminal visual act but doing what's called "News Adjacent Targeted Advertising." Here's just one example. The TV news production company INNX in San Diego produces TV news health features such as "The NyQuil News" hosted by former NBC News correspondent Lucky Severson, for 200 NBC stations that's watched by about 9 million consumers a day. And what they do is let advertisers select from among its upcoming segments those which best match their products, such as NyQuil, to figure out where a promo for NyQuil could be attached to a news story about the flu. So, you just thought you were getting some non-biased news story about the flu season, when in reality you got a carefully designed Subliminal Commercial to manipulate you into buying their flu product! It's all working together hand in hand, even in the news!

    And the reason why they are doing this is because the ad breaks are getting more and more crowded and competitive and people are buying more and more Personal Video Recorders like TiVo or have DVR capabilities in their cable subscriptions, which allows them to fast forward or skip outright all the commercials. And they can't have that! So, these new stealthy Subliminal "branding" techniques give the advertisers the means to get around these technologies enabling us to skip over their traditional commercials and still be manipulated into buying their products by simply inserting them "inside" the shows. We are being "forced" to watch their commercials no matter what! Why? Because they work! But that's not all.

The **2ⁿᵈ disguise** they use in Stealth Marketing is called **Digital Branding**.

And these are the digitally created images that do not exist in the real world but look very real on your TV screen. For instance, fans watching Major League Baseball on ESPN will see what seems to be product billboards on the walls behind home plate. But fans at the game will not – they don't exist. ESPN has a deal with digital ad leader Princeton Video Image (PVI) to insert what's called "Virtual Ads" into the network's game broadcasts. In fact, pretty much all of the sports broadcasts around the world are now using this Subliminal "Virtual Ad" technique on their viewers, whether people realize it or not, as the transcript of this video shows.

This clip shows a soccer game being played in the stadium, but as the film is running slower than real time you can see an ad come on the screen for Wild Turkey Bourbon. The players can't see it because it's not printed on the field like the other ads are. As they continue to cross the field another ad is brought onto the screen, also for Wild Turkey. And again, farther down the field there is another.

At another soccer game the ads that are shown on the screen are for Victoria Bitter. Again, this isn't anything the players can see, it is all subliminal. As they go a little farther across the field another ad for Home Timber & Hardware is shown over the players.

As people are at the beach and playing in the water the image of Kellogg is shown printed across the sand.[7]

What a bunch of fakeries! All designed to Subliminally get you to buy stuff and/or influence your choices and behaviors! But don't worry. They'll never use this on us! Yeah, right! Anybody starting to see a pattern here?

But that's not all. It's not just the Sports Industry getting in on this Subliminal "visual" technique. "Virtual Ads" are also being embedded into Television coverage of events and things like the arrivals of the Grammy Awards. There they displayed a "Virtual Street Banner" for Denny's, a "Virtual Entry Canopy Logo" for Candie's, and "Virtual Logos" on sidewalks for Ford and MasterCard. TV viewers saw the ads but arriving celebrities did not. Why? Because it was all digitally superimposed on the screen! We cannot escape this stuff!

In fact, sometimes, these "Virtual Advertisements" are digitally "flashed" on the screen even in music videos. The following video is from a rap group called Jovanotti featuring their song, "Penso Positivo" which means "positive task." Here you will see them doing a parody of "product placement" in a grocery store with the name of the song "Penso Positivo" on several products. But is that the only product being shown? See if you can catch the subliminal one, they digitally tucked in.

The video starts out with a dog running through the aisles of a grocery story. As people try to catch it to put it outside, the dog suddenly turns into a lead singer of the band playing in the market. The message that keeps coming into view is for Penso Positivo. This is what seems to be the message in this song that they are singing as well. As far as the human eye is able to see there are no subliminal placements.[8]

Here's a photo on the left that was digitally flashed on the screen for an actual popular European detergent called Dixan, right towards the end of that clip. Now, the irony is, the music video was making a parody of product placement, yet they subliminally tucked in an actual product placement!

And not so surprisingly, others in the media industry are also using this same Subliminal technique in kid's videos. Once again, here's just one example from Disney. It's a clip from their cartoon, "Oliver & Company."

A clip from Oliver and Company: As the dogs are running through the town they are going up and down the streets and the signs that they pass show the different ads. As you are watching them run and yell to get away from their pursuer your mind catches the pictures they pass.[9]

Remember, this is just a tiny clip of one cartoon. First, they set the pace for this Commercial Branding with an advertisement for COCA COLA right out in the opening credits!

Then, on the next page, you see an ad for RYDER trucks tucked away in this dark alley scene.

And then, just in case you forgot to think about it already in your brain, here's another advertisement for COKE.

And then here's an advertisement for GMC vehicles. Apparently for the adults who are stuck watching this cartoon with their kids.

And just in case you forgot about it again, here is yet another ad for COCA COLA.

Hey kids, apparently, they want to influence you now to think about growing up and "PLAYING LOTTO" as was flashed here.

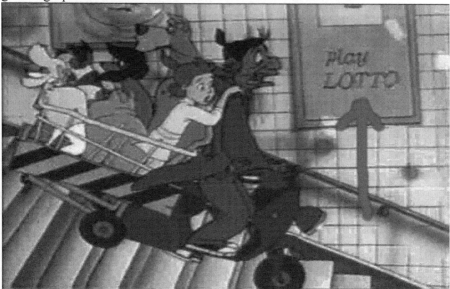

And then the final scene is one huge last gasp for advertisements as you see here:

with ads tucked in for SONY, TAB, USA TODAY, and yup you guessed it, in case you forgot again apparently, COCA COLA!

And believe it or not folks, they even use this Subliminal "visual" technique in Magazines as well. Nothing escapes their grasp when it comes to "visual" media. In fact, let me give you just one quick, easy, sneaky and deceptive example! Here we see a seemingly innocent photo of Marilyn Monroe, in a magazine, right?

Or is it? A close up reveals the truth.

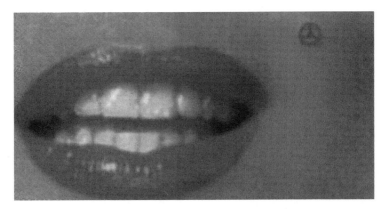

As you will see the mole on her face is actually a digitally enhanced picture of a Mercedes logo.

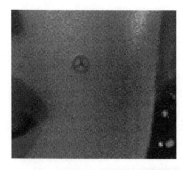

Talk about Subliminal Seduction!

And so, the question is, "Does all this visual Subliminal Stealth Marketing really work on us?" You bet it does! As we saw in the videos, Reese's Pieces was a relatively unknown candy. But after it appeared in the movie *ET* the sales leaped 66% in just three months. Then there's Red Stripe Beer. Red Stripe was a small import beer from Jamaica before it was seen in the movie *The Firm*. But once Tom Cruise took a swig, sales increased by 53%. And then, speaking of Tom Cruise, after he wore a pair of black Ray Ban sunglasses in the movie *Risky Business,* sales immediately tripled. Then there is the dress Demi Moore wore in *Indecent Proposal*, the backpack Julia Roberts carried in *The Pelican Brief* or the glasses Harrison Ford wore in *The Fugitive,* or the Subway sandwiches in the movie *Coneheads*, all of which caused a massive consumer outcry raking in millions upon millions of dollars for the advertisers. Pretty stealthy, isn't it? They know it works. That's why they use it!

In fact, sometimes they will even use this "visual" Subliminal Technology to "sell you" a politician or particular political party. Here's a classic clip of a Republican Presidential TV ad back in 2000 that two Democrat senators asked the FCC to review. Why? Because it was discovered they used Subliminal Technology in it! When the ads slowed down, the word "RATS" clearly appear on the screen in large, white letters, sending an ominous message about their opponents.

RNC Campaign ad: *"Under Clinton/Gore prescription drug prices have skyrocketed and nothing has been done. George Bush has a plan. Add a prescription drug benefit to Medicare."*

George Bush: *"Every senior will have access to prescription drug benefits."*

Narrator: *"And Al Gore opposed Bipartisan Reform. He's pushing a big government plan that will allow Washington Bureaucrats to interfere with what doctors prescribe. With the Gore prescription plan, bureaucrats decide. The Bush prescription plan, seniors choose."*[10]

Here is the freeze frame of the "message" they wanted you to Subliminally "think" about the Democrat party.

They are a bunch of "RATS" aren't they? This is why the two Democrat senators filed their complaint with the FCC because the opposing party really was using Subliminal techniques to influence voters.

Now, I personally think that both Democrats and Republicans are a bunch of RATS, as was exposed in the last election cycle. But be that as it may, anyone who uses this Subliminal Technology to "visually" influence others without their knowledge is a RAT! Including in this manner. Speaking of "flashing" a Subliminal message. The following are some freeze frame shots of Subliminally flashed ads that appeared on the Sci-Fi Channel. The actual ads appeared for only a fraction of a second and were placed between the regular commercials and the shows. Here's a photo number one of Liberty Medical services that flashed for a fraction of a second, 9 minutes and 2 seconds into the program, "Buck Rogers into the 21st Century."

Then this is the Dell Computers ad that flashed for a fraction of a second, 18 minutes, 47 seconds into The Twilight Zone program.

And on the following page is a photo featuring the product Swiffer flashed for a fraction of a second, 45 minutes, 38 seconds into the show Stargate SG-1.

And finally, here's a photo featuring Progressive Insurance that flashed for a fraction of a second, 1 hour, 54 minutes into the program called, Stargate Atlantis.

In fact. This Subliminal Flashing technique has become so commonplace that these particular gentlemen have made a hobby out of it, seeing how many times they can catch them doing it to us.

Newscaster #1: *"It comes as no surprise to learn that we get a lot of complaints about ads on television. They are too loud, too offensive, or just too annoying. I love them because they pay my wages."*

Newscaster #2: *"Whether they are good or bad, at least you are aware they are happening."*

Newscaster #1: *"Now hold on, what about ads that are too short, you blink, and you missed it. Could there actually be a case of the dreaded Subliminal imaging? Our reporter on the scene investigates."*

Reporter: *"Peter McDermott is of sound mind and body, but lately he says he has been seeing things. At first these messages were too quick for Peter to see."* He begins his interview.

"How did you notice it?"

Peter McDermott: *"Well, to be honest, my son noticed it. His eyes are younger than mine."*

Reporter: *"Then Peter started looking for them."*

Peter McDermott: *"I saw ads for the Lego movie, for Captain Marvel, Continental soup and for chewing gum."*

Reporter: *"He has now realized that they have been there all along. And Peter doesn't like it. These Subliminal messages are aimed to get into our heads without us even realizing it. He asked us for our help, and we couldn't resist taking a look. Consciously that is."*

Peter McDermott: *"Something is just not right, and they shouldn't be there."*

Reporter: *"Peter watches all evening."*

Peter McDermott: *"I could probably see 3, 4 or 5 of these one frame adverts."*

Reporter: *"Most TV ads are about 15 seconds long. Maybe 5 if you are watching online. A fresh frame is about 14 milliseconds. Barely there, hardly worth worrying about."*

As they are sitting in front of the TV watching a program.

Peter McDermott: *"That was an angel."*

Reporter: *"All of this was happening on Sky TV on the History Channel. Plenty of Tutor England and ancient history but doesn't explain this phenomenon. Peter wasn't too fast until he saw an ad for liquor flash up on the screen."*

Peter McDermott: *"The one that really upset me was the range of alcohol, and I think that was not acceptable."*

Reporter: *"He felt it was unacceptable and might cause more of a risk to someone more vulnerable."*

Peter McDermott: *"I'm not a drinker actually, but you are quite right. There could be the potential that someone might be drying out or on the wagon and this could just be the little thing that tips them over."*

Reporter: *"But would it, actually? We need an expert. Someone that would offer a glimpse into the human psyche and the world of advertising."*

Dr. Lisa McNeill, Otago University Business School: *"I think this consumer actually has a point when he said this doesn't seem to be the right thing to do."*

Reporter: *"But that is because of the bad alcohol ad, whose ad is more regulated than most."*

Dr. Lisa McNeill: *"What you are talking about is an ad that probably sits around the threshold, between superluminal, are able to be observed and* the subliminal, not able to be observed. And that threshold differs by individual."

Reporter: *"The difference between subliminal or not comes down to how long or short the ad is or whether you are aware there really is an ad."*

Dr. Lisa McNeill: *"So, we know that these subliminal ads do have an impact."*

Reporter: *"Turns out there are lots of examples that sit just at our level of awareness."*

Dr. Lisa McNeill: *"Just walking down the street. You smell this smell of hot bread, and you walk into the nearest Subway. That is a really good example of how it is actually working for you. Or when you walk into a store on High Street and you react negatively to the music being played. That is a good cue to you subconsciously that that store just isn't for you or hair products that really don't suit who you are."*[11]

But go to another store that is playing music more suitable to your generation and for some reason you just want to stick around and shop more! Go figure! And again, notice how they admitted that not only "flashing" "visual" Subliminal messages on TV was really going on, but they even admitted to the Subliminal "smells" and "sounds" being used on us as well. Anyone starting to see a pattern here? Maybe I'm not that "crackpot" that some of you might have been thinking about me at the beginning! Now you know why we've entitled this book, *"Subliminal Seduction: How the Mass Media Mesmerizes the Masses."* It's really going on, folks, on a multitude of levels whether we realize it or not, and it's wrong!

But that's still not all. Now we get to another immoral section. Not only is it Subliminally encouraging people to be greedy and materialistic immoral…

But the **2nd way** they visually subliminally seduce us is by **Sex Marketing**.

And as we saw in the beginning, this is one of the most commonplace "visual" Subliminal techniques they use on us to get us to "lust" after their products. This is why they have the phrase in the industry, "Sex Sells." It's not by chance. It really works and they use this "sex sells" technique on us in a variety of ways including subliminally.

First of all, I'm only going to give you just a couple of examples of this since we've already dealt with this a little bit in our previous chapter. But again, what I'm about to show you is, frankly, the tame stuff. I cannot, in good conscience, show you just how graphic they really get with this "visual" sexual manipulation. It's sick. But let's take a look at a few more samples of Subliminal Sex Marketing. Now, the word "sex" can be embedded upside down, airbrushed in, distorted, or even appear so commonplace that we don't initially perceive it altogether, yet our brain does, subliminally. For instance, here's one of the most common cited examples of a Subliminal picture showing a group of flowers.

How many of you can see the picture within that picture? Yeah, for those of you who can't see it right away, let me help you out.

As you can see the word "sex" is clearly embedded within this seemingly "innocent" photo.

Or how about this example we saw earlier. In 1990, Pepsi actually withdrew one of its "Cool Can" designs after someone protested that Pepsi was subliminally manipulating people. How? By designing the cans so that when the six-packs were stacked in grocery stores, the word "sex" would emerge from the seemingly "random" design.

Now, for those of you who still can't see it, let me help you out again. Check out the pictures, side by side, on the next page. See if you can see anything wrong with these seemingly innocent cans.

As you can see, the design on the can clearly spells the word "sex" when they are stacked together. Is this just a mere coincidence? I don't think so! Critics alleged that the red and blue lines on the can were far from being random, and I would agree!

Or, on the next page is a lady smoking a cigarette seemingly enjoying a nice summer day while the wind is "innocently" whipping through her hair.

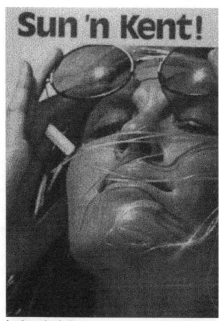

Or is that all that is in her hair?

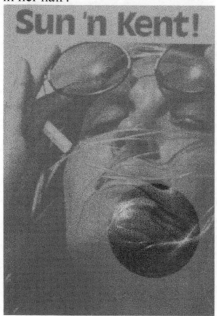

As you can see now, the word "sex" is clearly and carefully airbrushed into her hair! Gee, I wonder why?

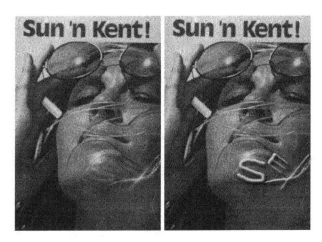

Or sometimes they skip the airbrushing of the word "sex" altogether and instead they subliminally "suggest" it in the picture itself. For instance, here's a seemingly innocent picture of adults enjoying a nice day at the beach.

Or is that all that's really going on here? Let's start from the left and work our way to the right and see just what's really going on here.

Now gee, what does that photo imply? Or how about this section below?

Or how about this?

And what's being implied here?

And gee, how about this suggestion?

And do you really think this pose is purely by chance? I don't think so!

And the point is this. Notice how many Subliminals were used in just that one picture!

Now, lest you think Hollywood is left out of this manipulative "visual" technique, here's the cover of their movie, "Silence of the Lambs."

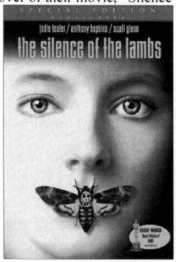

Now, I don't recommend you watch it, but check the cover out for yourself. If you notice, there's a butterfly on Jodi Foster's mouth.

But if you look even closer, you'll see a skull on the back of the butterfly. And if you keep on looking even closer, you'll start seeing something inside the skull.

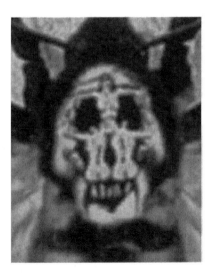

A close-up reveals the skull is actually a picture of a bare-breasted woman with outstretched arms with two other ladies kneeling before her with their naked backs exposed.

It's sick! And their feet in this position are actually what appears to be the skull's teeth. In fact, the supposed outline of the skull is actually a circle made up of other naked women as well. Gee, why would they do this?

In fact, as was already demonstrated, one of the most popular ways they "sexually" Subliminally seduce someone "visually" is by airbrushing the word "sex" right into a photo as you can see various styles of it on the next page.

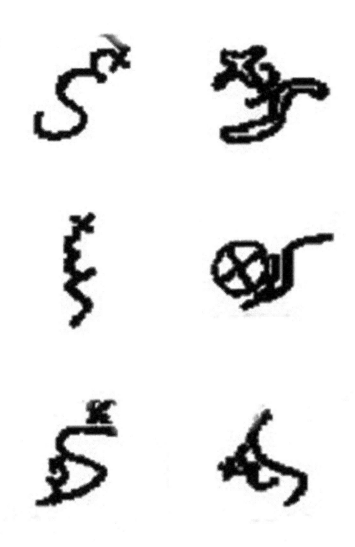

As you can see here, it could be upside down, backwards, squiggly, or freehand but it doesn't matter because our eyes still pick up on it and so does our brain.

For instance, here is a seemingly innocent looking Ritz Cracker. Or is there more than meets the eye here?

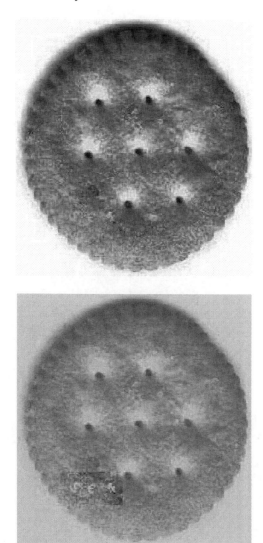

Of course, there is! As you can see, carefully airbrushed into the grain of the cracker is the word "sex." Why would they do that?

Oh, but crackers aren't the only culprits they do this to, so does alcohol. They're big on it! Here is a photo featuring an icy glass of Gilbey's Gin.

But is that all that's going on in that glass? Of course not! As you can see clearly, the word "sex" was airbrushed into the ice cubes on purpose!

And here's Camel's cigarette ad giving the command, "What you're looking for."

So just what is it that we are looking for? Well, apparently, it is the Camel's logo that just happens to be formed in this lady's hand from the dripping ice cube in her mouth.

And secondly, the other thing we're apparently looking for in her mouth is an ice cube with the word "sex" airbrushed right into it.

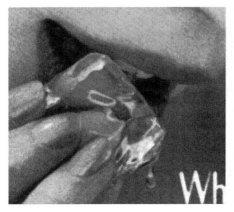 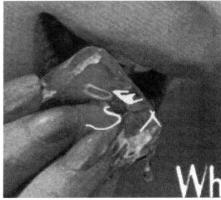

Is "sex" what I'm supposed to be looking for if I buy a pack of Camel cigarettes? That is what they're implying! Why?

Again, these are the extreme "tame" examples. I can't share with you the gross graphic examples that are also out there on a massive scale. It's again, gross just how far these people will go to subliminally seduce us into buying their products with Sex Marketing!

And here's the point. Think about it. No wonder people today are having a problem with wickedness and immorality like the Bible said would happen in the Last Days. No wonder our society is getting more and more perverse. No wonder people often say, "I can't get these wicked immoral thoughts out of my head. They won't stop! Why? Why? Why?" Well, maybe it's because of all the media you're ingesting is subliminally bombarding you with those wicked evil thoughts! Think about it! Shut it off and watch it go down!

And I say that because common sense tells us that people on average today watch 28 hours a week of television with this stuff on it, whether they realize it or not, and if only half of those 1,095,000 advertisements that hit us every year are sexual in nature, that means we are being sexually assaulted nearly 550,000 times every single year. And this is legal? No wonder things are going down the tubes like the Bible warned about! No wonder people are having a hard time with "sexual" temptations! In fact, every once in a while, you'll even get some so-called experts who will actually admit in public, that these types of "visual" Subliminal Manipulations are really being done to the public, as seen in this interview.

UWTV Classics, Upon Reflection 1993 – Anthony Greenwald, Psychology Professor, University of Washington: *"A lot of psychologists believe today that the great majority of information processing, processing stimulated from the environment sounds, sights, everything, is done unconsciously and the conscious is the top level which is only brought into play for the really important things that require action at the moment."*

Interviewer: *"Do you think that with how busy our environment is, how many messages there are, how much is going on, that we can only pay attention, I mean common sense tells you that you can only be conscious of a very small fraction of what is actually happening?"*

Anthony Greenwald: *"Right, well, we can only use an analogy coming over the television air ways, through the cable wires or whatever, are many stations, and many sources of information. We focus on one at a time because the tuner of our TV set will only pick up one at a time. But the way our minds function is as if we are monitoring all of them, so it might be as if we are looking at a TV screen that had all of the channels on it simultaneously, on separate pieces of the screen, separate channels, and whatever drew our attention, would blow up and command the whole screen, and focus on it, and that would be our conscious attention."*

Interviewer: *"Now, in advertising, these subliminal messages have been used deliberately, and we have a number of examples that we are going to take a look at now. Where some messages are being sent, we may not even know it was there because a lot of them, that we are going to look at most often, here we have Pepsi."*

Anthony Greenwald: *"Yeah, we have got a series of images here and I picked them out to illustrate a number of points. When we are talking about advertising, what we mean by Subliminal, means various things. Most people believe by Subliminal; it means something very briefly on or very dim. The examples we are going to be looking at are ones we can see clearly and there is Subliminal in another sense which is that they are not the focus of our attention. They might be processed unconsciously. Now this example comes from a can design that Pepsi used a few years ago. I don't know how easy it is to see on the screen. But one can see when these cans are rotated properly and stacked in a certain way, one can see the letters "sex" on the can on the right side."*

Interviewer: *"That is a big Subliminal message. "sex" shows up a lot."*

Anthony Greenwald: *"Who knows what kind of affect it has. Now here we have another case* (an ad for Salem Cigarettes) *unlike the Pepsi can, that we were looking at before where you see the letters "sex", this one is where people may disagree whether the letters are there, but if you look at the red ribbons on the tree, they, in their surrounding could look like the script, "sex." This makes it more obvious, but I have to say that it has been air brushed onto this slide and it has white dashes to outline the letters so the viewers can see where they are. Those white dashes weren't in the bigger one.*

So, I started with this because I think it is the popular view of what Subliminal advertising is toward "sex" on Ritz Crackers, ice cubes or whatever."

Interviewer: *"Now this is from a popular film."*

Anthony Greenwald: *"This is a different type of thing. Here is a very nice graphic work which is the poster from 'Silence of the Lambs' and you see that bee in the lower left. It's worked into the model's mouth, so it looks like a very interesting graphic thing. But when we go close up in that we see something else. We will see something that would be visible to someone that is looking at this poster closely, but you would really have to* *focus on it. Here we see on the back of the bee is something that is clearly a skull, there is no doubt about that. But what viewers may not be able to see is that this skull is actually composed of three or four human figures that appear to be nude. Can you see that? It may not come across that well. The eyes of this skull are the backs of heads and the bones that look like they are coming down from those eyes are the torsos. But this is another example of what is sometimes meant by Subliminal.*

Now, this is sort of continuing the buried head theme in the image. This is a jewelry advertisement, and we are looking at it in a full shot now. But in the top center there is one piece of jewelry, a ring appears, this is a cover of a catalog, and if you take a closeup of that what you see is (the inset is not in the ad) *that small piece of jewelry in the top center that has been air brushed and doctored by the artist to contain what looks like a vampire's head to look like the one in the inset."*

Interviewer: *"Now let me ask you, here is this picture with daisies and is all springy, why would they do that? What is the purpose of putting that message in what seems to be a very contradictory context?"*

Anthony Greenwald: *"It's conceivable that this image, as part of the larger cover shot, had some subtle value of triggering an association, and why associate jewelry with this evil, well maybe the vampire is regarded as seductive, maybe it's an association with seductiveness.*

Here is another one that is even more direct than the last one. The question is, why is this here? This is an ad for clothing (children's clothing). From this full view, you may be able to see in the tree in the back of one of the models. It looks like two eyes, a nose and a head, maybe an animal head. If we look at it closer, it actually looks like a wolf's head. One might say maybe that tree trunk just happened to look like that but almost certain it was no accident. It isn't a natural shape, people who do this kind of commercial work are very conscious of all the detail. This was, I am positive, put there deliberately. So back to your question, why? Why would this be put there deliberately?

Here we have something else. This is the only shot of this one that we are going to see (an advertisement for Dior). *This is a men's clothing ad. You will notice in the background a very vague shape that you could probably guess is a very evil looking head. It might be an animal, or it might be a skull or something like that. It is sinister. And the way the whole ad is photographed with this male model being in the dark, suggests darkness and sinister. It looks like some sort of a menacing animal.*

Here is one for perfume (Nuance by Cody). *This one to me is a very interesting one, because now just like the last clothing ad, this gets into the domain where those things that are off in the periphery may be having some effect on the person because of the association to trigger and can help to sell the product. Now from this distance you can't tell what's there but if you look at a closeup you see that on the walls of the buildings in the background, and you really have to look closely, are a few words, one of them is 'fresh' and the other word is 'soft' and in this particular image there were several other words on different buildings.*

You can see these if you look all over the surface of these, they are in plain sight, but the typical person is going to go by them rapidly, looking through a magazine, would not dwell on them. Nevertheless, the eye would pass over them at the periphery, and they could register their association value and in this case their association is soft and fresh and a few other things to capture the image of the perfume.

Here is our last image (Camel Cigarettes). *People might think we are looking at it because of the cool camel who has been regarded as a rather objectionable character and perhaps phallic image, I won't get into that aspect of it, but there are some other interesting aspects of this ad. At the periphery of our attention and our attention is very focused on that character but look at what is happening off to the side with the things we are not paying much attention to."*

Interviewer: *"Like war is being waged in the Persian Gulf?"*

Anthony Greenwald: *"He's wearing the Tom Cruise sunglasses from 'Top Gun' and we have a plane taking off from the carrier deck."*

Interviewer: *"And you have this woman leaning in one side of the picture."*

Anthony Greenwald: *"So this is capturing the idea of the hero in that movie, 'Top Gun' and it's not at all blatant. It's very subtle. But I think it is the kind of thing that can register to get our attention."*[12]

In other words, it really does have a Subliminal effect on us, whether we know it or not. But that's not all! They get even trickier than that!

The **5th way** Subliminal Technology is being used on us is **By Our Sense of Psyche**.

Now, it's bad enough as we've already seen that we are being subliminally manipulated by our sense of smell, our sense of savor, our sense of sound, and even our sense of sight, but there seems to be no end in sight as to the means of which these advertisers and corporations and media moguls will go to seduce us into buying their products, or what the Bible calls idolatry. And believe it or not, they are now going straight into our brains or psyche in order to manipulate us into becoming unrestrained consumers with 100 percent accuracy. I know that sounds freaky and something straight out of George Orwell's 1984, but it's really being done, and I want to expose that to you. Now, for those of you who think they could never really manipulate our brains that easy, I'm going to show you just a few common techniques out there showing us how easy it really is to do.

The **1st way** to easily manipulate a person's brain is **With Color**.

Pay attention to all the different ways we can be manipulated simply by using "color" variations or a combination of both. It happens whether you realize it or not as these researchers admit.

Color Affects our Moods

"Color, direction, quantity, and intensity of light strongly affect our moods, mental abilities, and general well-being. California architect Vincent Palmer has experimented with color and intensity of interior light, and he has found that he can modify the behavior of his guests by changing the color of the light around them. Light quality affects people's emotions and physical comfort, thereby changing the volume and intensity of their conversations and even the length of their visits."

Color Affects Our Emotions

"Light can be used to elicit an emotional reaction in the viewer. Lighter values, being brighter, seem less serious or threatening. It can evoke a sense of calm, quiet, or warmness. On the other hand, other light values can suggest drama, excitement, or even conflict. In the same way, overall darkness may provide feelings of sadness, depression, or even mystery. Yellows, oranges, and reds give us an instinctive feeling of warmth and evoke warm, happy, cheerful reactions. Cooler blues and greens are automatically associated with quieter, less outgoing feelings and can express melancholy or depression."

Color Affects our Minds

"Color, a component of light, affects us directly by modifying our thoughts, moods, actions, and even our health. Psychologists, as well as designers of schools, offices, hospitals, and prisons, have acknowledged that colors can affect work habits and mental conditions."

Color Affects our Bodies

"People surrounded by expanses of solid orange or red for long periods often experience nervousness and increased blood pressure. In contrast, some blues have a calming effect, causing blood pressure, pulse and activity rates to drop below normal levels. The affect may be purely psychological, but the results are very real. Perhaps you have read of the

workers in an office painted blue complaining of the chill and actually getting colds. The problem was solved not by raising the thermostat, but by repainting the office in warm tones of brown."

But that's not all. The **2nd way** to easily manipulate a person's brain is **With Illusions**.

Now, these examples are pretty wild, or should I say "mind-bending." This first one is called the Muller-Lyer Illusion.

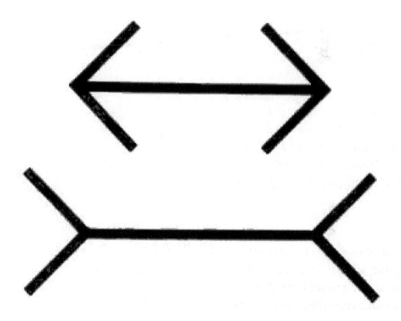

Take a look at the picture and tell me which one of the two lines are longer than the other. Did you say the bottom one? Wrong! If you measure them, you will find out they're exactly the same length. Our brains see the lines as different because we have been "taught" to use specific shapes and angles to tell us about an object's size. But it is all an illusion.

Or how about this example on the next page. It's called the Ponzo Illusion.

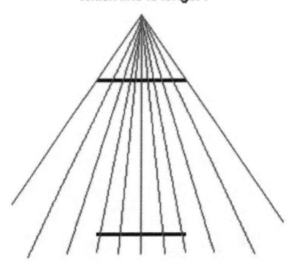

Now, take a look at this picture and tell me which line is longer? It's obvious, the top one, right? Wrong again!

If you look really close, you'll see they are actually the same length as well. The addition of lines creates the "illusion" to our brains that the top line is longer when it's not.

Now, the following example on the next page is called the Elephant Illusion.

Take a look at this elephant and count how many legs it has. Piece of cake, right? The more you look at it the harder it is to count the legs, isn't it?

Or how about this one. The Straight-Line Illusion.

How many of these lines are straight? None of them, right? Wrong! If you look at each one carefully, you'll see they're all straight!

Or how about the Construction Illusion.

Take a long look at this picture and see if you can tell which side is the top, and which side is the bottom. Not as easy as it looks, is it? Not at all!

And this one's called the Hermann Grid Illusion.

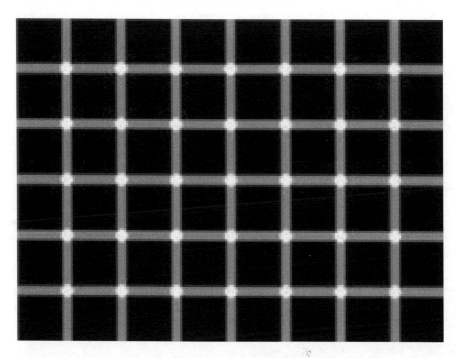

Take a look at this picture and see how many black dots you can find. I'll give you a couple of seconds. As you noticed, for some reason, as soon as you "see" the black dots, they are "seen" to disappear, don't they? It's wild, isn't it?

Or, how about this one. It's called the Rotating Illusion.

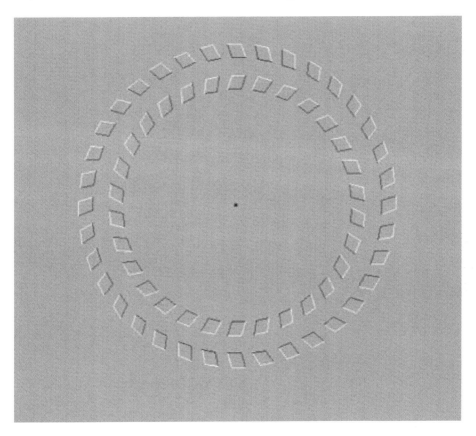

Now, concentrate on the dot in the center of this picture and focus on it. Then start to move your head in closer and then back again and "see" what happens. I'll give you a few seconds. It's trippy, isn't it?

The circles appear to rotate, don't they?

And then we have the Pulsating Illusion.

Stare at this photo for as long as you can and you'll not only "see" that it "appears" to pulsate, but it actually starts to give you a headache!

And finally, we have the After Image Illusion.

Now, stare at the four vertical dots in the center of this picture for 30 seconds. Now, close your eyes and tilt your head back, and you will soon see a circle of light.

Keep looking into this circle and tell me what you "see."

For those of you who stuck it out, you actually "saw" an image of Jesus, didn't you?

This is called the After Image Illusion.

Now, these are just a few examples of how easy it is, whether we realize it or not, to manipulate our brains. And the reason why I bring this up is because these same "brain illusions" are being used on us in commercials! Shocker! Let me show you the proof!

Color Contrasting Commercials

A while back there was a Dodge commercial that featured a camera shot that immersed the viewer into a bright green forest and it proceeded to just "fly around" in this seemingly innocuous shot of constant green foliage.

But, right at the last second, the screen changed to a bright red Dodge Ram symbol on a black background, specifically.

Now the question is, "Why would they do that and what does that have to do with a Dodge vehicle? Here's the answer. And I quote:

"Another phenomenon that defines complimentary color is the afterimage affect. Certain color pairings are difficult to look at and our eye experiences a conflict in trying to perceive them simultaneously. The effect is strongest with reds opposed to blues and greens."

So, this seemingly "innocent" Dodge commercial was creating an "after image" in our minds of their Dodge Ram logo in our brains by first having us focus on a "green" background then flashing quickly to a "red" Dodge logo! It is not by chance folks! It is all carefully scripted to manipulate us! They know exactly what they are doing! Here is another example.

Motion Contrasting Commercials

Now, whether you realize it or not, another Subliminal technique they use on our brains to manipulate us into buying their stuff with commercials, is by showing us people or objects that freeze in mid-motion and then continue in regular speed shortly thereafter. Much like they did in the movie, *The Matrix*.

Now the question is, "What does that have to do with a product and why would advertisers do that to us?" Well, again, here's the answer. And I quote:

"Startle effects are useful in promoting products as well, because the fast-paced influx of images moves faster than we can react, leaving us no time to reflect on the information. People may remember the name of the product or brand, but the details of the ad itself float in one ear and out the other."

So, in other words, this next Subliminal brain technique "startles" our minds in order to embed their product into our minds! They even admit it! Here's another one.

Flashing Contrasting Commercials

Some of you may recall some of the commercials they use on us that specifically hit us with rapid fire multi-colored flashing words or objects much like a strobe light effect. It is annoying and frankly, sometimes, it's not easy on the eyes. So, the question is, "Why are they doing this?" Well, it might be annoying and irritable to our eyes, but it has positive effects for what they are trying to accomplish, subliminally. Here's the proof:

"The right brain is stimulated by the movement of animations, flashing lights, and brightness. Listening to the dialog or music of a television commercial while special effects such as flashing, are perceived by the subconscious mind. And these experiences are stored into the memory banks of the right cerebrum."

So now they are hitting our brains with these bright flashing lights in order to subliminally store a product into the memory of our brains! Anybody starting to see a pattern here? So much for nothing to worry about. Subliminal Technology is just a big hoax! They would never use it on you. Yeah right! They are using every trick in the bag on us, and it's wrong! In fact, and as I mentioned before, it's now considered a "science."

The **3rd way** they manipulate our brains is **With Neuro-Marketing**.

Now, as we saw earlier, these same advertisers and corporations, and who knows who else, are not only using Subliminal Techniques including the color and brain illusions, as we just saw. But they are also using the latest and greatest Subliminal trick of all time called "Neuro-Marketing."

And what it means is that they are now working at manipulating us right down to the "neuron level" in our brains and looking for what they themselves have coined, the proverbial "buy button." I'm not joking! And this is the part of the brain that they say can not only be Subliminally "triggered" to just "entice us" to buy their product, but believe it or not, to "force us" into buying their product. Don't believe me? Let's take a look at that proof.

First of all, if you feel like advertisers get into your head now, you ain't seen nothing yet. They are literally going straight down to the brain level to manipulate us and there's big bucks in it! The technology to monitor and alter brain waves dates back to the 1970s with actual patents for apparatuses for remotely monitoring and altering our brain waves. Most of the earlier techniques relied upon some kind of calibrated flickering but was easily lost in airway broadcasts. What was missing was a reliable mass media delivery system.

Now thanks to digital technology, they've got the delivery system solved! And since they've now got their reliable mass delivery system, to do their dirty deeds, they're now getting even more specific with their manipulative methods.

The latest trends in advertising and marketing are now using a technology called "Neuroscience" to look way beyond just how to influence our choices and to now directly affecting our brains at a physical level. It has given birth to what's called "Neuromarketing." In fact, BrightHouse Neurostrategies Group launched the first Neuromarketing company back in 2002 and promised in a news release to, "Unlock the consumer mind."

And what they do is use Medical Brain Imaging machines such as MRI to map people's brain responses to certain exposures to products, advertisements, packaging, commercials, and even movie trailers to test their effectiveness and specific response. And what they've learned is that the act of deciding whether to make a purchase lasts 2.5 seconds. And when the possibility of buying something first occurs to a person, the visual cortex, in the back of the head springs into action. Then a few fractions of a second later, the mind begins to turn the product over, as though it were looking at it from all sides, which triggers memory circuits in the left inferotemporal cortex, just above and forward of the left ear.

(See how specific they're getting?) Then, when a product registers as a "strongly preferred choice" – the goal of every advertiser – the action switches to the right parietal cortex, above and slightly behind the right ear. And if a spot near the top of the brain dubbed "the magic spot" is fired, a consumer is no longer deliberating, they're "itching" to buy. Literally, at that point, they are saying to themselves, "I'm going to do it. I want it."

Now, this is what they are learning by mapping our brains with this Neuroscience, and it has created a whole new field of science called, "Neuromarketing." It's all admittedly moving toward their goal of finding the "buy button" inside our skull, and to test products, packaging and advertising and their ability to activate it, this "buy button." And listen to this. Not just to boost sales for consumer products or merely to generate fads, but even to win votes for political candidates. Another pattern we're starting to see, isn't it?

And BrightHouse Neurostrategies has lured in big corporations as customers who are paying big bucks for this kind of research, like the Metropolitan Museum of Art in New York, Home Depot, Hitachi, Coca-Cola, Delta Airlines, Georgia-Pacific, Metlife, just to name a few. And they've done so with the blunt purpose, "Closing the gap between business and science – with the goal of getting people to behave the way corporations want us to." Can you believe that? Straight out of the horse's mouth! Or as one of BrightHouse's executive put it:

"What it really does is give unprecedented insight into the consumer mind. And it will actually result in higher product sales or in brand preference or in getting customers to behave the way they want them to behave."

Total manipulation folks! And they have the audacity to even admit it, including in this video.

Advertisment: *"My name is Marie, and I am a liberated woman. I live my life as I wish. Which is why I have chosen this brand. This skin cream (Cash/Desir) I can't do without it. The sensation, when I apply it, is a real* *pleasure. When I get home from the gym, I often stop at this burger restaurant (Cash/Burger). I love its smell and the toasted bread. And it is so nice inside. I reserve my plane tickets through this travel agency. (Cash/Voyages) There are so many images there that maybe I will want to make a return trip to the sunshine. It is much like my bank. I recently changed over to this credit company (Cash/Credit) and took out a loan for what I need to buy for the kids. My name is Marie, and I'm a liberated woman. I like these products because I'm worth it."*

Narrator: *"Well, Marie, do you really believe you are a free and liberated woman? Did you know that the smell of fast food is deliberately designed to produce an emotional reaction? The travel pictures are deliberately chosen to arouse your desires. The bank slogan is especially created to play on your fears. And your body cream has been developed to spark a desire within your brain. You didn't realize that of course. Why should you? It's all calculated by the latest technology from neuroscience, like an MRI or an electroencephalogram. These new sales methods are a technique to control your brain. It's called neuromarketing. But my dear consumers, you are not meant to know about it."*[13]

Can you believe the audacity there? In other words, they're hiding it from the average Joe, but they're really doing it! Straight out of the horse's mouth again! I wish I was making it up, but we've got to deal with the facts. In fact, when people do deal with the facts, they come back with statements like this. One guy said:

"It sounds like something that could have happened in the former Soviet Union, for purposes of behavior control."

You think? Another critic said this:

"We already have epidemics of obesity, diabetes, alcoholism, gambling and smoking – all tied to marketing."

In other words, keep up all these Subliminal Techniques and now add in all these specific manipulative techniques gained from

Neuromarketing and it's going to make these things worse! In fact, one person said just that:

"Any increase in the effectiveness of advertising can be devastating to the public."

In other words, you keep this up, it's going to destroy our society! And that's exactly what God said would happen when you're living in the Last Days! Yet, believe it or not, some people actually support these manipulation methods on people. One lady actually stated this:

"If I feel a little bit crummy or a little bit down, my fallback strategy is shopping. I think if they can find a way to help us find a way into that magic little feeling that shopping can give you — if you do it right and you get the right thing and you don't spend too much money, hats off to them. Thank you. I think it's a service."[14]

Wow! Talk about brainwashed! I guess this Neuroscience is really working! But as you can see, we've been lied to about this Subliminal Technology for years, literally for decades. It's the same ol' tired lie they use on us! "Don't worry. You can trust us. This is all just a big hoax. They would never use it on you." Yeah, right! When the whole time, they are doing it, out there in plain sight, for anyone to see, or should I say, for those who have eyes that "want" to see. I think some people "don't" want to deal with this truth. They want to go back to their dream world and put their head in the sand, but that doesn't help anything! And so, for those of you who might still be in denial even after all you've seen so far, I will first of all, close this chapter by inviting you to listen to some logic.

Ask yourself, "Why would advertisers spend $163 billion dollars every year on these manipulative techniques unless of course they really do work?" Either it works for them, or they're the dumbest people on the planet who have tons of cash to burn! Trust me, common sense tells you these guys repeatedly spend $163 billion dollars every year because they know it works, and they are going to get a whole lot more than just $163 billion back! They're in it for profit! Plus, that's just the top 200 advertising companies spending that much. Who knows what the real total spending amount is from all the companies worldwide?

Secondly, I'd like to invite you to listen to these kids. If you don't want to listen to logic, then at least listen to the honesty of these kids who share with us the harmful effects the mass media really has on them and how it really influences their behavior.

- *Cigarette billboards tell you to smoke, so do beer commercials on TV. Too many advertisements for alcohol. They can make you buy toys and make you buy cigarettes and beer. I feel that they want me to smoke or maybe drink. I don't like the way they portray alcohol with men and women, especially women.*

- *Bad language. Saying bad words. False language. A radio speaker announcer says bad words, and TV can make people do bad things. I can't watch TV without hearing a swear word and can't watch TV without seeing violence. The "Simpson's" have bad language and do bad stuff. On KDWB Tone E. Fly is sick! Swearing in commercials. Songs on the radio like on The Edge and 93.X.*

- *There are too many killings on TV. There is way too much violence on TV and causes me to think that most people are like that, and the people that they show who are violent are scrubby and dirty and causes me to think all violent people are like that. TV shows violence and gory things a lot. They talk about Strep-A and meningitis and killings and it gets me worried. Violence on TV. Too much violence and too much doing drugs. It scares me. It influences me to have bad nightmares.*

- *It says everyone wants to be skinny, so I want to be skinny. Lies and bad pictures. Too much sex. It makes me think getting hurt is cool. Sex on TV is extremely bad. It causes people to think everybody is doing it. TV makes you feel that you have to be beautiful in order to be a good person.*

- *Guns and violence. The media is big on stuff like violence, bad examples. A lot of violence and killing. Too many drugs and killing. It has caused me to be more violent in some of my actions. All the killing and drugs. Fights. They give ideas about making bombs and other things. There's a lot of violence. They show shootings. Video games that have too much fighting. Watching people getting beaten up. To get into fights. I see crime a lot, guns, kidnapping, and gangs. The media is always showing people dying. It has changed me to be not so nice because I saw it on TV.*

- *Violence and sex. Drugs and sex. Show bad stuff like 1/2 pornographic. All of the sex and violence. X-rated shows on cable. Women as sex objects. See other people naked on TV all the time.*

- *It has caused me to swear more often. Lots of foul language. Swearing on radio and TV.*

- *It causes me to have a bad attitude toward my parents.*

- *It says do what you want to do and don't listen. It has made me make bad choices and do something I thought was right that is really wrong. Commercials make you buy things. Candy. I think they are trying to get you to buy something. We can be influenced by choices people on TV make. It gives some people false images. It draws conclusions about people that might not be true.*[15]

How does that old saying go? "Out of the mouths of babes." When are we going to wake up and realize that there really is a Subliminal Seduction going on and that the mass media is mesmerizing the minds of the masses, for some pretty nefarious purposes, I might add. And we are just getting started!

So now let's turn to the unfortunate news that these people and their dastardly deeds are about to get even worse. As those kids admitted, they were also being influenced by "other" forms of media out there as well, not just Television. It included radio, billboards, print media and the like. So, could it be that "all forms" of media are "telling a vision" for us to accept that is really manipulating us for another nefarious purpose? Unfortunately, the answer is yes as we will soon discover in our next chapter.

Chapter Three

The Manipulation of Newspapers

As if we haven't seen enough proof already, there really is a Subliminal Seduction going on with the mass media that is mesmerizing the minds of the masses. Just as I finished the research on the last section, the following just happened to come my way. I don't think it's by chance. Not only is it from the current time of this research, but as you will see once again, it reiterates, in very blunt terms, that Subliminal Seductive Technology really does exist. It really is being used to manipulate us right up to this very day! Here's what it stated.

"How Subliminal Images Impact Your Brain and Behavior. Subliminal messaging – we have all heard about it. But it doesn't really work, right? New research from the University of Texas at Houston suggests that Subliminal images 'can' change our brain activity and behavior.

What does Subliminal messaging entail? Subliminal messages are words or images presented below our conscious awareness. Usually, we think of short frames cut into a video feed, where the Subliminal message appears so quickly (Usually less than one-tenth of a second!) that our minds do not register their appearance. On the other hand, Supraliminal messages are presented for longer periods of time, such that we can consciously see them.

So, does Subliminal messaging actually affect us? There is evidence dating all the way back to the 1960s, which suggests that showing Subliminal images 'improves' behavioral performance. And recent studies show that Subliminal imagery 'can' significantly alter neural activity and behavior. This influence is probably why Subliminal messaging in advertising is banned in many countries.

Subliminal messages live in our everyday lives, where we may never notice them. While behavioral studies over the years have suggested Subliminal messaging can modulate our choices, this new study shows how these behavioral changes occur on a single neuron and network level.

Now you know – Subliminal messaging is not a myth! Maybe the next time you pick up your favorite drink, take a moment to ask yourself, 'Why 'is' this my favorite brand?'"[1]

Yeah, why "did" you choose that one? There it is again folks, right out in the open. Subliminal Technology and usage of Subliminal Messages is not a conspiracy theory! We really "are" being bombarded with them and "manipulated" to this day, and for some pretty nefarious purposes I might add.

Yet, it's about to get even worse! And I say that because not only are corporations and businesses and governments and even politicians using this Manipulative Technology to "sway our minds" all over the world, but they are also using various forms of media to "control our minds" around the world! And they're doing that by "controlling" the information that we receive through various media outlets. It's not fair and balanced. It's not an "open, honest and free exchange of ideas" that you and I, the public, can trust as reliable and dependable information. Not even close! Rather, whether you realize it or not, the people behind our modern-day media are literally "controlling the narrative" of what is being broadcast. In essence, "we only get" what they "want us" to get. We "only think" what they "want us" to think. Which, by the way, is a form of "mind control."

And frankly, it's led to this kind of society that I'm about to show you depicted in this clip. It's from a movie called, "They Live." And it's the classic scene where the lead character finds a box of "special" sunglasses that enables him to "see" what's really going on in his society. Let's take a "look."

The opening clip of this movie, "They Live" shows the main character with a box full of sunglasses. He is searching through this box in front of the overflowing garbage cans. This box of glasses seems to have been put out with the trash. As he searches through the box, he finds a pair and puts them in his pocket. He looks around to see if anyone noticed him and then closes the box and puts the box of glasses in the trash can and covers it with paper.

As he proceeds to walk down the street, he puts the glasses on but as he sees the ground through the glasses he stops in amazement. Something is not right. He takes the glasses off and looks around the street. He puts the glasses back on, and the sign reads "OBEY." But when he takes them off again the sign has a normal advertisement. He tries it again. He puts the glasses on and looks at the sign, and it again reads "OBEY." This is weird. He takes the glasses off and places his hand over his eyes. This cannot be real.

He puts the glasses back on and looks at a different sign that is advertising a trip to the Caribbean, but this sign reads in bold print "MARRY AND REPRODUCE." He walks a little farther down the street and comes to a Men's Apparel Shop. Even though the sign reads the name of the store, when he puts on the glasses the same sign reads, "NO INDEPENDENT THOUGHT" and a smaller sign in the window reads, "CONSUME." Without the glasses the same little sign read "CLOSE OUT SALE."

Trying to figure out what is going on he looks down the street, puts the glasses back on and all the signs are different. In bold print they are telling the public "WORK 8 HOURS", "SLEEP 8 HOURS", "PLAY 8 HOURS", "WATCH TV", "BUY", "SUBMIT", "CONFORM", and "STAY ASLEEP." He then walks past a magazine stand and again there are signs saying, "WATCH TV" and "BUY" and "STAY ASLEEP" and "SUBMIT" and "NO THOUGHT."

He picks up a magazine and looks inside. All the same words are in there except for one additional statement. "DO NOT QUESTION AUTHORITY," When he takes the glasses off and looks at the pages again, they are normal articles. While he is flipping through the pages, he put the glasses back on and a man steps up, to also look at the magazines. As he looks up at the man, he is shocked to see an alien being staring back at him and the alien asks, "What's your problem?" He realizes he is staring at this alien, so he takes the glasses off and the alien is now a normal human dressed in a nice business suit. The man in the suit repeats, "I said, what's your problem?" As the man walks off, he pays the clerk for the magazine and then turns back to look at the man with the glasses. Looking through the glasses, this man is now an alien again. The alien takes his change and walks to his car and leaves.

In unbelief he stands there in shock. The clerk comes over to him and says, "Hey Buddy, you going to pay for that or what?" The clerk is normal looking, but the money in his hand has printed on it "THIS IS YOUR GOD." The clerk continues, "Listen Buddy, I don't want no hassle today. Either pay for it or put it back." He puts the magazine back and turns to walk away.

When he walks to the corner, waiting for the light to say walk, the word "Sleep" keeps being repeated over and over. He walks past a hair salon and looks in the window. There are 4 or 5 ladies sitting under driers but when he puts the glasses back on, one of them is an alien. The other ladies do not even realize that they are talking to an alien. There is a lady loading her car and talking to two other ladies about a Lamaze class but as he puts the glasses on, he sees that the lady talking about the class is also an alien. His legs can hardly hold him up as he walks into a little grocery store. Half of the people in the store are aliens. He takes a deep breath as he passes a human man talking about his feelings to an alien being.

Then the TV comes on and the newscaster speaking is also an alien. Behind him is the word, in large, bold print, "OBEY." The newscaster is saying, "We have faith in our leaders, we are optimistic to what comes of it all. It really boils down to our ability...."[2]

Can you imagine living in the world like that? Where everybody was being subliminally manipulated behind the scenes through the media. And no one had any clue that they were subliminally being told to obey, never question anything, consume, marry, reproduce, and go back to sleep! Can you imagine what a sick reality that would be? I mean, good thing it was just a movie! Unfortunately, that really is the world we live in right now!

And this study, whether you realize it or not, are the "glasses" that we are trying to get you to "put on" so you can "see" what is really going on behind the scenes, just like that guy! In fact, just like in that movie, there really are a cadre of people around the world who are responsible for creating this subliminally controlled society. Only they are not "Aliens," but rich "Elites," who do get all the good high paying jobs and positions, and of course, they get to enjoy the lavish lifestyles, while the rest of us slumber around. And these "Elites" have the same goals in mind like that movie portrayed. That is, total "Subliminal" mind control and domination of every "non-elite" person on the planet!

I know it sounds crazy, but once you realize this, it not only starts to explain why our society is in the shape it's in, but it also starts to reveal yet another sign from the Bible that we are living in the Last Days. Let's take a look at that one.

2 Peter 3:3-7 "Above all, you must understand that in the last days scoffers will come, scoffing, and following their own evil desires. They will say, 'Where is this 'coming' He promised? Ever since our ancestors died, everything goes on as it has since the beginning of creation.' But they deliberately forget that long ago by God's word the heavens came into being and the earth was formed out of water and by water. By these waters also the world of that time was deluged and destroyed. By the same word, the present heavens and earth are reserved for fire, being kept for the day of judgment and destruction of the ungodly."

So here we see the Bible gives another sign that we are living in the Last Days. It not only says that we would see a rise of wickedness like never before, but also like we saw in the first sign earlier. But now, here we see there would also be a rise in the "scoffing" toward the things of God like never before. And this "scoffing" is specifically towards the truths of God. Including the fact that God really did judge this planet once with a global flood because of the previous rise of wickedness. And now it says He's going to do it again, because of another future rise of wickedness. The first time it was with water, the second time will be by fire. These "truths" are what the Last Days society will be "scoffing" about.

But hey, good thing we don't see any signs of our world "scoffing" at the truths of God today, including His next coming judgment! Yeah, right! We're living in a world where "scoffing" at God, the Bible, Jesus and His soon coming return, are at an all-time high! Here's my point. How did this happen? And in such a short amount of time. Well, believe it or not, it's due to the rise of this Manipulative Subliminal Technology that we're being subjected to all across the planet via the media. The timing of its "manipulative" arrival is not by chance! It's just in time to fulfill yet another Bible Prophecy sign that we're living in the Last Days!

You see, as I mentioned at the beginning, this Subliminal Manipulative Technology is being used not only to "sway our minds" with Subliminal Messaging that "tells" our subconscious what to "buy" and "consume" just like in that movie. But it is also being used to "control" or "instruct" our minds "to go back to sleep," "to not question authority," "to obey" whatever they, the "Elites," say, including, believe it or not, "scoffing" at the things of God.

So, let's "keep our glasses on" and let's see in this chapter how the mass media system, that's built and owned not by "Aliens" but rich "Elites" really are "controlling the narrative" so they can Subliminally "control" and "instruct" our minds with "only" the information they "want us" to get.

The **1st way** the global media is "controlling the narrative" to "control our minds" is with **Newspapers**. You see, for the longest time, when it came to sharing information, mankind was severely limited, let alone having any real ability to become a global information sharing society like we are today. For many centuries, if a person wanted to share information, it was not only local, but they would also use the tedious method of cuneiform tablets, which is a communication method of slowly pressing shapes into clay tablets. Still others would use other primitive methods like writing on reed parchments or animal skins, or even broken pieces of pottery to get their message across. Century after century, this was about the extent of man's communication methods. Until now! All of a sudden, not that long ago, a serious breakthrough occurred. It was called the Gutenberg Press and it radically changed, overnight, the means of which, the method and the rapidity of information sharing, as this video transcript shows.

James Meigs, Editor-in-Chief, Popular Mechanics: *"Everybody knows the story of Gutenberg and his printing press, nevertheless it is one of the greatest stories in the history of inventions. The ability to produce books is severely limited because to produce a book a person must sit with one book and painstakingly copy it into a second version. They were fantastically expensive, they took a long time to make, and these limiting factors kept them from being distributed to people.*

Narrator #2: *"Gutenberg saw an opportunity. He knew it was possible to print in a new way that it would make it less expensive. He saw it as a tremendous money-making opportunity.*

Prof. S. James Gates Jr. University of Maryland: *"Gutenberg had the idea of removable type.*

James Meigs: *"Removable type where he didn't have to make new ones for each page. You just had to rearrange the letters on the pages that you want."*

Narrator #2: *"Once movable type came into play and the production costs went down, knowledge flourished. The ability to have access to this world of knowledge was absolutely transformational for Europe."*

James Meigs: *"The notion of pressing blocks against paper to make images or texts did really come from China but no one ever turned it into a manufacturing process. That was the difference. He could make thousands of them, quickly and efficiently. He started to see the end of the artisan era and the beginning of the manufacturing era. If you think about the impact that had, it's really hard to underestimate it."*

Prof. S. James Gates Jr., University of Maryland: *"Knowledge up to that point was the property of only a small number of people, rather it be royalty or a religious order."*[3]

In other words, back then there were "Elites" "controlling" the information people received, but with the invention of Gutenberg's Press, it was wrenched from their hands, for a time anyway. Because as we will soon see, these "Elites" have not only grabbed that informational control back, but they have done it on a massive global level this time!

But as you can see, with the invention of the Gutenberg Press, information sharing went from the slow hand-written "artisan" method, to the fast-paced "manufacturing" method. Now we could crank out information like never before. And it wasn't just books either. Gutenberg's Press also gave rise to "mini-informational books" or what we call, "Newspapers." With these new forms of media, people could now find out the "news" of what was going on in their "local" world in a very inexpensive way and on an industrial scale.

Soon the first weekly printed newspaper appeared in Europe in 1605 called, "The Relation" and from there it took off like wildfire! Soon it hopped the pond to America with America's first daily newspaper in 1784. And then, one of the early American newspapers was called the "Publick Occurrences Both Forreign and Domestick."

So, as you can see, from early on, the trend of this new information sharing technology called Newspapers was not only communicating "local" information to people, but "global" events as well. It was a huge breakthrough, especially if your goal was to "manipulate the minds of the masses" by "controlling the narrative" of information that people receive! And of course, as we will soon see, that is exactly what's going on with newspapers today, literally, all over the planet!

But this is why today we have an estimated 18,000 newspapers in circulation in about 102 countries, all available right now at our fingertips, any time, instantly, connecting all of us, thanks to another invention called the Internet. We'll get to that information technology in one of our final sections. But when it comes to newspapers, the top 10 Global Newspapers right now are:

- The New York Times (U.S.)
- The Guardian (Great Britain)
- The Washington Post (U.S.)
- The Daily Mail (Great Britain)
- Kompas (Indonesia)
- Liberty Times (Taiwan)
- USA Today (U.S.)
- The Wall Street Journal (U.S.)
- The Daily Telegraph (Great Britain)
- China Daily (China)

In fact, let's take a look at the "massive global scale" this largest of all global newspapers operate on. When it comes to "The New York

Times," it is light years ahead of the Gutenberg Press! Look at this video transcription!

The Insider Reports: *"This ten-mile-long roll of newsprint paper alone will soon turn into 30,000 copies of the New York Times, that will be on the newsstands and doorsteps of thousands of people in about five hours. It is 10 pm at the College Point, neighborhood in Queens, New York, and the shift has just started for the workers at this printing plant who will meticulously work to the early morning until 3 am, to print one of the most respected journalistic publications in the world. That paper you read in the morning, over your cup of coffee, probably will have come from one of these 27 different printing plants, passed through hundreds of hands, and then inspected by thousands of pairs of eyes before landing in your hands. We visited this 300,000 square foot printing facility in Queens to find out all that goes into this process."*

Mike Conners, Managing Director of Production Dept.: *"Getting the paper out is an equation. We are graded every day on our arrival time. We have 52 trucks that go out most nights. If they are on time, we get a 100%, if they are late, it decrements from there. We are measured every morning on how our arrival times are."*

Insider: *"First a digital copy of the newspaper is sent to the printing plant. This paper is the final version where the editors, writers, and the copy editors sign off on to be printed. From there each page of the newspaper is digitally transferred to a plate using a laser machine. A plate is a sheet of aluminum that contains the image of the newspaper page on it. Each plate is equivalent to a single page of the newspaper. The plates are made in a room made with special yellow lighting to help protect the plates unit page from exposure. This is what will eventually transfer onto paper. But we will get to more of that later. Plates that are ready to go to printing are stored in groupings based on the section they will be in, in the actual newspaper."*

Mike Conners: *"This is the Sunday Arts or Drama as we call it in our world." He tells us as he points to the different pages ready to go to printing.*

Insider: *"This helps keep the plates organized and lets everyone know where each plate belongs in the printing stage. While the printing of the plates is taking place, giant rolls of paper are be transported by clamp trucks which help the rolls move around. These rolls of newsprint are stacked and stored in a large warehouse. Each roll is 10 miles long and makes 30,000 newspapers. When the rolls are ready to be used, the outside paper is removed by hand. The paper is now ready to be brought over to the press machines. But before the printing happens the damaged parts of the paper roll must be removed. Wrinkled or damaged paper cannot be used for printing, so that portion of the roll is recycled.*

Now this is where the true magic happens. The printing of the paper. The plates are individually put into the press cylinder by hand. They are purposefully connected to one another, so they print in the direction the newspaper reads. And of course, there would be no newspaper without the ink. To print that many newspapers in a single night, massive amounts of ink are stored in containers and transported to each printing press through metal pipes. As the cylinders begin turning, ink get splashed on to the muted image on the plates and then that image gets transferred to a sheet of printing paper creating the physical copy of the newspaper. Although the process needs to be quick and efficient, the printing plant doesn't sacrifice quality.

It's about 3 am at this point of the process. The paper is checked as it comes down the assembly line. Employees will look at the alignment of the paper and the coloring, making sure all the imagery has transferred correctly. A large ruler that can accommodate the dimensions of the paper is used to check the positioning of the columns of each page. Once the papers are ready to go, they are wrapped up and shipped out to be placed on the doorstep of their readers by early morning. Although the Times is printed across the country, the Queens printing plant produces nearly

41% of the publication's daily papers. Although almost 80,000 copies of the paper are printed every hour, timing is everything."[4]

Yeah, I would say so! Gutenberg eat your heart out! What a long way it has come from the 1400s. Now they are cranking out newspapers with all kinds of information for people to digest, local to global, and that's just 1 out of 18,000 of these kinds of operations going on all over the world, every day!

Here is the "assumption" with it all. Gee, I sure hope these rich "Elite" owners of these newspapers are being "open and honest and fair" with all this information we are digesting from them! I mean, surely these "Elites" wouldn't use these now "global" forms of media information sharing to subliminally "manipulate" the minds of the masses in order to create a desired outcome for their own nefarious purposes, would they?

Oh, how I wish I could say no to that, but then I would be lying like these rich "Elitists" are. Folks, I'm here to tell you, and as you will see in a moment, the usage of newspapers to subliminally manipulate the minds of the masses for some "Elitist's" nefarious purposes has been going on for quite some time. In fact, one doesn't have to look very far back in history for the obvious proof.

It is now a well-established fact that one of these rich "Elitists" who used his newspaper for his own nefarious purposes was a man named, William Randolph Hearst. And it is an historical fact that he clearly used all the newspapers he purchased and amassed to manipulate people, even here in the United States, on a massive scale for "his" chosen desired outcome.

Narrator: *"Hearst traces its origins to March 4, 1887, when 23-year-old William Randolph Hearst placed his name on the masthead of the San Francisco Examiner with the title of proprietor. The Examiner had been owned by William's father, George Hearst, a California rancher, and miner whose interests tended to lean more towards politics than newspapers. After George Hearst was elected as a US senator in 1886, he* *turned the newspaper over to his son. The younger Hearst assumed the responsibilities as both editor and publisher and quickly transformed the state Examiner into what he called the monarch of the dailies.*

He bought the most advanced printing equipment of the day. He dramatically redesigned the newspapers look and hired the best journalists he could find (Example: Jack London and Mark Twain). *Within a few years the new Examiner was a runaway success and would point to a new era of American journalism. Hearst purchased a second newspaper from the east coast, The New York Journal, in 1895. Soon he would start newspapers in Chicago, Los Angeles, and Boston. Hearst's newspapers pioneered innovation such as multiple-colored presses, wire syndication, and the first color comic sections. By the 1920s Hearst had 28 newspapers nationwide read by one out of every four Americans"*[5]

Wow! One man, a rich "Elite" was responsible for "feeding information" to 1/4th of the whole United States, by simply buying up a whole bunch of newspapers! Good thing that's not going on today! Yeah right. We will see it is being done on a massive "Global" scale in just a minute.

But think about the "power" that this one, rich "Elite" had over the minds of the masses. Surely, he reported the news in all his newspapers around the country in an "open, honest and fair manner." He wouldn't use them to "manipulate" people for his own nefarious purposes, would he? Unfortunately, the answer is yes, and this is yet another well-documented historical fact. William Randolph Hearst used all his newspapers to "manipulate" the minds of the masses by "controlling the narrative" in those newspapers, to get "a desired outcome" "his outcome" from the people.

For instance, it is now known that Hearst used his newspapers to, "Whip up popular support for U.S. military adventurism in Cuba, Puerto Rico and the Philippines." He also used his newspapers to, "Distort world events and deliberately try to discredit his critics." And it is even reported that, "Hearst's' papers accepted payments from abroad to slant the news."

So much for being "open, honest and fair" with journalism in newspapers! In fact, they even gave a term for the "manipulative" tactics Hearst used in his newspapers. It's called, "Yellow Journalism." And as you will soon see, it still continues to this day! This is how powerful a "manipulative" tool newspapers have become!

Paperboy: *"Read all about it!"*

Narrator: *"In 1898, William Randolph Hearst and Joseph Pulitzer were fighting a circulation battle on the streets of New York. Hearst decided he was going to scoop Pulitzer by reporting on the escalation of the Cuban war for independence and he was going to do it with style. He staged a daring rescue of Evangelina Cisneros, a popular Cuban prisoner. Hearst then delighted the public with the sizzling story of her escape. Her rescue, however, wasn't the only thing that Hearst had staged. The war in Cuba was little more than bar fights and minor skirmishes. Hearst ordered his reporters to exaggerate the severity of the conflict to make it more interesting. I guess boring reality sells fewer papers.*

Because these early escapades were printed with yellow ink, an innovation at the time, this attention grabbing, and deceptive reporting was given the name Yellow Journalism. A term still used today to describe tabloid trash and dishonest journalism.

Even documentaries have a sordid history. In 1922 film maker, Robert Flaherty made a feature link documentary called 'Nanook of the North.' In it he chronicled the struggles of an Inuit family as they tried to survive the harsh Arctic environment. Here they are building an igloo to keep safe from the bitter winds. And here they are listening to a record for the first time. Hey, don't eat that! (One of them is seen putting the record into their mouth.) *Again, there was just one problem. It was all staged. In fact, the family shown in 'Nanook of the North' owned their very own gramophone player regularly enjoyed listening to music on it. But according to Flaherty, showing the family as backwards and primitive told a much more interesting story, and would sell more tickets at the box office.*

Today the media relies on sensationalism more than ever. While newscasters like Dan Rather and Brian Williams have been fired for telling outright lies, others have found success while focusing on tantalizing news stories over ones with substance. It seems like we are living in an age where it is becoming increasingly difficult to find useful and reliable news."[6]

In other words, the stuff that Hearst did with his newspapers way back in his day, is still going on today, on an even more massive scale! "Yellow journalism" or the new modern term for this old-fashioned historical "manipulative" practice, "fake news" is everywhere! Rich "Elitists" really are using newspapers, still to this day, to "manipulate" the "minds of the masses" to "generate their own desired outcome," including in the area of "politics." Believe it or not, with newspapers, they want to make sure that you "only "vote for" who they "want you to vote for." Why? Because their "pre-chosen" candidates will allow these same "Elitists" to maintain their power.

Remember, it was "politics" that Hearst's dad was really interested in. He was the one who originally gave William Randolph Hearst his first newspaper back in his day in San Francisco. In fact, for those of you who might find this hard to believe, here is just a small sample of proof that this "political" manipulation with newspapers is still going on today!

Fox News Reports: *"Now we have it over on the big board, coverage is 92% negative,* (negative coverage of Trump Administration by ABC, CBS, NBC Nightly News) *so perhaps that does shade the manner in which some of the mainstream networks do cover this administration and politics in general."*

Guest: *"The thing that got the most coverage was actually the Russian investigation, which if you look at the polling the people care about the least."*

Fox Reports: *"According to the media research center between June and September, the Russian collusion investigation on the networks had 342* *minutes and immigration close to 300 as well. But the administration's economic achievements got 14 minutes. Not much."*

Guest: *"No, not much at all. The things that affect the lives of millions and millions of people and it deserves a lot more scrutiny."*

Fox Reports: *"Absolutely, and it deserves more publicity so people will know what is going on."*[7]

Yeah, what a concept, that the News would actually report the "truth" and "true events" instead of "fictitious" biased narratives like what Hearst did with his newspapers in order to "control the narrative" and "manipulate" people. In fact, people in the news industry today admit that these "Elitists" really are using their news outlets to "control" what people "believe" in order to "generate" "their own desired outcome."

Matthew Gentzkow, Richard O. Ryan Professor of Economics and Neuhauer Family Faculty Fellow: *"What drives the choice of slant by media? You can think of two very different narratives. One, the slant is driven by the political preferences of the owners of the news outlets, maybe by the editors or reporters. The other slant is chosen just to maximize profits and therefore is driven by whatever the readers of the newspaper want to hear. So, there is a correlation out there that everybody can see, and I think it is one of these cases that people get a little mixed up between correlation and causation and we show that causation is ultimately coming from the economic side. But in fact, you see that most conservative things are run by mostly conservatives and liberal things are run mostly by liberals."*[8]

In other words, whoever owns that newspaper, and whatever "mindset" they have, they are going to push that "mindset" onto other people, whether they realize it or not. It's not an "open, honest and fair" market of information sharing like you would think! It's clearly slanted and biased in order to get people to "think" what these rich "Elitists" "want you to think" just like Hearst did back in his day with politics. Which is why one person stated this:

"There was a time when the media, at least made an attempt to be fair to both sides in the political debate. Sadly, that time is now dead and gone. And it shows in the contempt many Americans now have for the media. Even so, don't shed a tear for the media's seriously tarnished reputation: They've brought it on themselves."

And sure enough, just like Hearst, these "Aliens" I mean, small group of "Elitists" have actually purchased up virtually all newspapers around the world and are now using them right now to "control the narrative" not only in the United States, but literally across all developed countries to "manipulate the minds of the masses" as well on a global scale. Check out the proof for yourself!

Billionaires love buying newspapers, even though the industry is struggling. Salesforce CEO Marc Benioff is buying Time Magazine for $190 million. Benioff and his wife Lynne say it is a family investment, calling themselves "caretakers." Biotech billionaire Patrick Soon-Shiong bought the LA Times for $500 million in 2018.

Soon-Shiong: (His goal is to preserve integrity of the LA Times). *"I believe there has to be a trusted news source, someone you could go to that is fairly impartial, that really wants to report the news."*

In 2013, Amazon founder & CEO Jeff Bezos bought the Washington Post for $250 million dollars. Bezos says buying the paper was a business endeavor, not philanthropy. The company made a profit in 2016 and 2017.

Jeff Bezos: *"I think a lot of us believe that democracy dies in darkness, that certain institutions have a very important role in making sure that there is light, and I think the Washington Post has a seat, an important seat because we happen to be located here in the capital city of the United States of America."*

Red Sox owner John Henry paid $70 million for the Boston Globe in 2013.

John Henry, Red Sox Owner: *"I invested in the Globe because I believe deeply in the future of this great community, and the Globe should play a vital role in determining that future."*

Casino magnate Sheldon Adelson secretly bought the Las Vegas Review-Journal in 2015. Former reporters and editors say he's tried to exert control over stories. But Adelson denies meddling in the editorial process.

Warren Buffet's Berkshire Hathaway owns more than 30 local papers.

Warren Buffet: *"I still love newspapers. I'm very glad we own them."*[9]

     Yeah, I bet you are! But don't worry! As the one billionaire stated, "We're only doing this so we can create a trusted news source that's fair and balanced." Yeah, sure you are. We know differently! Reports are now coming out that, *"15 Billionaires own America's news media companies."* And *"It's sparking a conversation about the ultra-wealthy's role in 'controlling' the news."* That's their words, not mine! They even go on to question if this trend is, *"A new way of using money to influence the media,"* and that, *"Billionaires have long exerted influence on the news simply by owning U.S. media outlets."* You know, like what Hearst did!

     Nothing new, folks! No need to reinvent the wheel here. It worked in the past for William Randolph Hearst, why not use it on people today. And they are! But as you can see, there really is a small group of "Aliens" I mean "Elitists" who are buying up newspapers across the United States in order to "control the narrative" to "manipulate the minds" of the masses for their own nefarious purposes. I'm not making this up! And it is not just the U.S. billionaires doing this, it's billionaires around the world also buying up newspapers in other countries as well!

Aaron Bastani: *"Billionaires control the media, and it is undermining democracy."*

Newscaster @ BBCPapers: *"This is a great set of papers; I've had such fun out there reading those. Anyone who thinks politics are boring would have their mind changed by these papers."*

Aaron Bastani: *"The Sun, Times, Daily Mail, Metro and Telegraph are all owned by individuals who are incredibly wealthy and don't even pay taxes in the UK. There's Rupert Murdoch who founded News Corp, whose UK arm, News UK, owns the Sun and Times stables. He lives in the US where he is a citizen. There's Viscount Rothermere with the Daily Mail, Mail on Sunday, and Metro. He lives in France and is a non-dom.*

And there are the Barclay Brothers with the Telegraph and Sunday Telegraph, not to mention the Spectator, twins so cartoonishly bad they have a secret lair in the Channel Islands. Murdoch leads the pack in wealth and is worth around $20 billion. The Barclay Brothers are worth 8 billion pounds, while the fourth Viscount Rothermere is worth a comparatively measly 1 billion pounds.

Rothermere, like his father before him, is based in France with his business interests operating through a complex arrangement of offshore holdings and trusts. The Barclay Brothers operate not just from their private island but also Monaco. Like Rothermere they too are non-doms although they claim that is for health reasons. Ever thought about maybe just trying a gym membership? All of these individuals use their papers to ensure their vested interests as a wealthy elite are best served. For example: ensuring any potential government is committed not only to lower taxes for the rich but also isn't serious about tax avoidance.

What is more, they lobby for privatization and outsourcing to always be preferred, no matter the cost or the catalogue of failure. Why? Because both act as a machine to take money from ordinary people and give it to the rich. So, every time you walk past the newspaper stand in the supermarket or at the petrol station, remember it is this handful of billionaires whose opinions you are staring at. It's like that stupid Facebook post your uncle sometimes makes except rather than ignoring it once a month you have to see it everywhere, every day. All these media barons have backed the Tories in recent elections not only through endorsements from their papers but personally. Rothermere favored Cameron and knew him, while Murdoch entered 10 Downing Street literally through the back door following the 2010 election when David Cameron wished to thank him for his support.

BBC Reports: *"Why did you enter the backdoor at number 10 when you visited the prime minister following the last general election?"*

Murdoch: *"Because I was asked to..."*

Aaron Bastani: *"More than just supporting the Tories in the past the Barclay Brothers have even donated them money. Guys, it is not supposed to be that obvious. Now we are supposed to believe that these moneybags do not interfere in what their papers say, but they do. Harold Evans, a former editor at the Sunday Times spoke of how he was often rebuked for 'not doing what he (Murdoch) wants in political terms' and how the two almost had 'fisticuffs' after he published an article by the economist James Tobin which Murdoch disagreed with.*

David Yelland, former editor at the Sun, made a similar admission but went even further, saying 'Most Murdoch editors wake up in the morning, switch on the radio, hear that something has happened and think, what would Rupert think about this – you look at the world through Rupert's eyes.' Now they don't own these papers out of the goodness of their hearts, these are ultra-wealthy, powerful individuals who own newspapers, which often make no money, because they want to influence political outcomes in a decisive way. This is one of the most insidious ways of undermining democratic politics."[10]

As well as "manipulating" people by "controlling the narrative" via the Newspapers for your "own" desired outcome! This is going on all over the world, whether you realize it or not! In fact, the terms they use to describe this global manipulation trend by these Billionaire Elites are, *"Concentration of Media Ownership, Media Consolidation, Media Convergence or even Media Oligopoly."* They all describe it as the

"Process whereby progressively fewer individuals control increasing shares of the mass media," and that this, *"Media ownership is generally something very close to the 'complete state control' over information in direct or indirect ways."* That is about as blunt as you can get folks! Their words, not mine!

In fact, these "Aliens" I mean small group of billionaire "Elites" are not only buying up the newspapers around the world to "control" what people around the world "think" about "politics," but just like the Bible said would happen in the Last Days, they are also using these same global newspapers to "control" people's "attitude" towards the things of God around the world as well. And wonder of wonders, it is turning them into a bunch of "scoffers" as well. Let's take a look a small sampling of that proof as well!

"Anti-Christian Media Bias is Real. I've done considerable work as a sociologist documenting anti-Christian bias in academia. There is plenty to be found. It seemed unlikely the academic world would be the only place such bias would arise, so I followed that up with research on the media and Christianity.

These findings show that many in the media are closed to the idea that Christians can face unfair hatred. However, conservative Christians face animosity in the United States that is just as high as Muslims face, but the media still dismisses anti-Christian hatred.

My prior research shows that liberals and secular individuals are more likely to have anti-Christian hatred than others. Of course, It's well known (and researched) that media personnel are relatively likely to be politically liberal and secular. That means many of our respondents may have a certain level of anti-Christian hostility.

This explains much of why Christians cannot rely on the media to tell stories sympathetic to their causes. It may also explain why some media personnel argue that religious freedom is unimportant. This makes the task of Christians who deal with the media harder than those who advocate other causes. They are naturally going to write a less supportive story."

"Exposing the Times' anti-Christian bias. It's rare to see a major media outlet be so honest about its ideological bias. But yet there was the New York Times reporter Dan Levin on Twitter the other day, openly soliciting negative stories about Christian schools.

The New York Times' long history of prejudicial coverage of religious Christians should cement your skepticism about its intentions. For example, The New York Times was one of many outlets that negatively reported that Second Lady Karen Pence had recently begun teaching at a private school that adhered to the Christian doctrine of her church.

Editors at The Washington Post and other large outlets incredulously wondered how Christian schools that still embraced traditional social values could even 'happen' in contemporary American society."

"The British Bias Corporation. At the BBC, cultural biases are built-in and nurtured. Not only are those biases tolerated, but public criticism of the media is shrugged off as 'uninformed.'

Over time, a media outlet eventually declares its own preferences and biases. It is hard not to, when you are in the business of making public declarations, and then defending them. Patterns emerge. A course is set.

An analysis of coverage demystifies the interests of media decision-makers and reveals prejudices, predispositions, attitudes and perspectives. Story decisions—what to pursue, what to ignore, how to play the news—reflect deeper issues, like the sense of responsibility surrounding the media institution itself—and the people who run it.

Bias against religion is itself a perspective that can manipulate culture and nurture elements that are harmful to the ideals of freedom of thought and freedom of faith.

Case in point: the British Broadcasting Corporation. The BBC's consistent attacks on religious freedom are a principal example. A review of the BBC's coverage of religion—overwhelmingly negative—provides an interesting map on which to mark their course.

The BBC has been widely criticized for its anti-Christian programming, including accusations of anti-Semitic bias in coverage of Israel and coverage that painted the Church as in decline, and irrelevant.

In a 2011 documentary, the corporation questioned whether Christianity had a future, describing the state of the religion as in 'terminal decline.' The public deserves better from its 'guardians' of truth."

"Media biased against Christians. Why does the news media promote Islam over Christianity? Why is Islam considered a great religion rather than a pagan religion? They force people to become Muslims with the sword rather than the love of Christianity.

In Saudi Arabia, the women are treated like furniture. Women cannot drive a car, and in their law can be beaten by their husbands.

In Turkey, in 1915, the Armenians were driven out of Turkey and slaughtered to the tune of almost 2 million people. There is a movie called '1915,' where at the end, it shows naked women on crosses all the way down the road in Turkey. This is how they treated the Christian Armenians in Turkey. Turkey will not admit it to this day.

The news media hates Christianity and will attack anyone who supports it. CNN, MSNBC, ABC, CBS, NBC all these are biased against Christianity."

And again, this is happening all over the world. So, as you can see, these "Aliens" I mean rich "Elitists" are buying up all the newspapers all over the world in order to "manipulate" the "minds of the masses" to produce the "behavior" and "political outcomes" "they so desire." Including, as you just saw, the "desire" to "manipulate" the "attitudes" of people, specifically against God and Christianity, so they will "scoff" at

Him on a global basis. This Media Manipulation has not only led to an "increase of wickedness" in people's behavior, like the Bible warned about in the coming Last Days society, but it has also produced a global society of "scoffers" towards the existence of God and His soon coming judgment! Total "manipulation" of the "minds of the masses," across the planet, just by simply "owning" all forms of informational media in this one outlet called newspapers.

Unfortunately, these "Aliens" I mean rich "Elitists" are not merely content with just owning all the newspapers around the world to "control the minds of the masses." In their insatiable lust for "power" and "domination" they are using another tactic that Hearst employed. That is, they are now branching out in "all different kinds of media" around the world and are buying them up too, in order to even more "control the minds of the masses" as seen in this next video transcript.

"Hearst also branched out into the burgeoning magazine industry launching a motor magazine in 1903. Over the next 10 years, he would acquire such legendary titles as Cosmopolitan, Good Housekeeping, and Harper's Bazar. In 1928 Hearst Magazines built its International Headquarters on 8th Avenue in New York City. In moving beyond the medium of print, Hearst soon began producing newsreels in partnership with a European film company. In 1929 Hearst formed a separate production company, Hearst Metrotone News to produce them worldwide. Hearst moved into radio broadcasting in the 1920s and in 1948 Hearst became the owner of one of the first television stations in the country, WBAL-TV in Baltimore. By the mid-century Hearst had built one of the world's largest media companies.

After William Randolph Hearst's death in 1951 the corporation forged ahead, continuing to diversify and expand its operations. In the 1980s Hearst became a founding partner in cable television's Lifetime and A&E Networks and soon also acquired a 20% interest in ESPN. In 1997 the corporation joined the leading ranks of a group of television broadcasters with the formation of Hearst Argyle Television. Hearst's magazines made news of its own in 2000 with its most successful launch in the company's history, O, the Oprah Magazine, a venture with the Harpo Entertainment Group. Today Hearst continues its founders enterprising spirit and is a leading global diversified media information and services company with more than 360 businesses. Each day at Hearst more than 20,000 people are at work helping to inform and entertain and inspire an audience of millions worldwide."[11]

Or, should we say, "manipulate" an audience of millions worldwide. As you can see, these billionaire "Elites" can't stop with just one media outlet to "control the minds of the masses." Oh no! They literally are out there amassing "all forms of media" in order to get the job done. Including radio, which will be the topic of our next chapter.

Believe it or not, radio stations and even the music they play on them, are being used to "control" people's "behavior" and their "minds" just like the newspapers do. You cannot escape their methods of control! They are everywhere! And as we will soon see, the media of radio and music are also telling us "to not think for ourselves," "to obey only what they want us to obey," "to never question them," and of course, "to scoff" specifically at the Bible, God, Jesus, and His Soon Coming judgment! Where have I heard that before?

I know it's hard to believe, but "keep those glasses on." Now let's begin this next journey exposing how this next form of "media" is also "mesmerizing the minds of the masses."

Chapter Four

The Manipulation of Radio & Music

The **2nd way** the global media is "controlling the narrative" to "control our minds" is with **Radio & Music**. You see, Newspapers are not the only Subliminal Manipulative Technology that's being used on us, on a global basis, by a small group of "Aliens" I mean rich "Elites" around the world. Believe it or not, Radio, including the music played on Radio, is as well. It too is out there "swaying our minds" subconsciously "telling" us to "buy" and "consume" and even "instruct" our minds "to go back to sleep," "never question authority," "obey" and "believe" whatever these rich "Elitists" "want us to believe" including the need to "scoff" at the things of God.

So, let's "keep those glasses on" and "see" how these next Media Technologies called Radio and Music are being used to "manipulate the minds of the masses" on a huge global scale. Frankly, as we saw with Newspapers, this "manipulation" that's being done first with Radio broadcasts, has been going on, as well, for quite some time now, whether people realize it or not. Let's "listen" to how it all began.

The History of Radio

"In the modern era, radio waves control everything. From the tunes in your car driving down the road to the police radio in the car that is pulling you over for not signaling your turn. These waves are undetectable and indivisible to human senses. But they make up the foundation of modern connected technology. The impact it had on our world is immense.

It all started in 1873 when James Clerk Maxwell showed mathematically that electromagnetic waves could travel through the air. Before long, in 1888, Heinrich Hertz demonstrated Maxwell's theory by demonstrating that someone could produce and detect electromagnetic radiation.

Then in 1892, Nikola Tesla used Maxwell's mathematical findings to demonstrate the sending and receiving of radio frequency energy. He proposed that this method could be used for sending and receiving information.

But it was Guglielmo Marconi who built the first wireless transmitter in 1896, which was capable of sending signals up to one and a half miles. He then proceeded to build the world's first Trans-Atlantic radio communications service between Clifden, Ireland and Newfoundland in 1901.

The next evolution in radio technology was the invention of the Spark-gap transmitter. This device allowed for the production of the first commercially available radio sets. But the Spark-gap radios were plagued with problems, mainly electrical interference. These were greatly improved with the invention and production of crystal radio sets.

The crystal sets were the first widely produced and widely used radio sets in America. These crystal sets were widely used in most American homes by the 1920s. It was the American family's connection to the rest of the country. This was fantastic, but there was still a problem that needed to be fixed. Up to this time all radios were being broadcast using AM waves.

AM stands for Amplitude Modulation. But the problem was that AM radio had a medium range and was prone to be blocked by urban structures.

But in 1933, Edwin H. Armstrong invented FM radio. FM stands for Frequency Modulation. This type of broadcasting method uses a wavelength that is less prone to static and blockage. It also has a longer range. With the invention of FM, radio needed only one more evolution to become what we know today.

That last advancement was the transistor radio. The transistor radio had particular advantages over the old crystal sets. The transistor sets were much cheaper than the crystal ones, used less power, had a smaller size and had a very long-life span."

"Over the last century radio has radically changed the course of humanity. And has rapidly accelerated the growth of the information age."[1]

As well as the growth, of yet another Technology, to "manipulate the minds of the masses." With Newspapers you could "read" what was going on around the world "local to global." But now with Radio, you can "hear" what is going on in the world "local to global." What an upgrade from Newspapers! Now you can "hear" voices, people, and the actual personalities! Why, it is almost like being there! A huge significant improvement, especially if your goal is to "manipulate the minds of the masses" across the whole planet! And so just like Newspapers, once this new "manipulative" technology called Radio sprung into existence, it too began to spread across the whole planet.

That is why today, we have an estimated 44,000 radio stations broadcasting throughout the world, 24 hours a day, 7 days a week, with over 15,000 of those Radio stations being in the U.S. alone. From the very beginning, Radio has proved to be yet another powerful form of Media to "manipulate the minds of the masses," including even causing death. Remember this historical broadcast? Let's "listen" in.

A Clip from the War of the Worlds Radio Broadcast

Music is playing when it is suddenly interrupted by an emergency alert.

Announcer: *"Now, nearer home, we come to you from Trenton, New Jersey. It is reported that at 8:50 pm a huge flaming object believed to be a meteorite, fell on a farm in the vicinity of Grover's Mill, New Jersey, twenty-two miles from Trenton. In the sky it was visible from a radius of several hundred miles. And the impact was heard as far north as Elizabeth. We have dispatched a special mobile unit to the scene, and we'll have our commentator Carl Phillips give you a word picture of the scene as soon as he can be reached there from Princeton. In the meantime, we take you to the Hotel Martinet in Brooklyn where Bobby Millette and his orchestra are offering a program of dance music."*

Orchestra music of 1938 is playing while they wait.

Announcer: *"We take you now to Grover's Mill, New Jersey."*

Carl Phillips: *"Ladies and Gentlemen, this is Carl Phillips again, out at the Wilmuth farm in Grover's Mill, New Jersey. I made the 11 miles from Princeton in ten minutes. I hardly know where to begin to explain your word picture of the strange scene before my eyes. It's something out of the modern Arabian Knights. I just got here, and I haven't had a chance to look around yet. I guess that's it! Yes! I guess that's the thing directly in front of me, half buried in a vast pit. Must have struck with terrific force. The ground is covered with splinters of a tree that it must have struck on its way down, but I can see the object itself doesn't look very much like a meteor. At least from the meteors I have seen, this looks more like a huge cylinder. What would you say Professor Pierson? What would you say the diameter of this is?"*

Professor Pierson: *"About thirty yards."*

Carl Phillips: *"About thirty yards and the metal on the sheath is, well, I've never seen anything like it. The color is sort of yellowish white.*

Curious spectators are now pressing close to the object in spite of the efforts of the police to keep them back. They are getting in front of my line of vision. Would you mind standing to one side, please? Oh, while the first man is pushing the crowd back, here's Mr. Wilmuth, owner of the farm here. He may have some interesting facts to add. Mr. Wilmuth, would you please tell the radio audience as much as you remember of this rather unusual visitor that stopped in your backyard? Step closer please. Ladies and gentlemen, this is Mr. Wilmuth."

Mr. Wilmuth: *"I was sitting listening to the radio."*

Carl Phillips: *"Louder, Mr. Wilmuth, louder please."*

Mr. Wilmuth: *"I was listening to the radio and kind of drowsy. That Professor fellow was talking about Mars, so I was half dosing and half..."*

Carl Phillips: *"Yes, Yes Mr. Wilmuth, then what happened?"*

Mr. Wilmuth: *"Well, as I was saying, I was listening to the radio, kind of halfway."*

Carl Phillips: *"Yes Mr. Wilmuth and then you saw something?"*

Mr. Wilmuth: *"Well, not first off. I heard something."*

Carl Phillips: *"And what did you hear?"*

Mr. Wilmuth: *"A hissing sound like this, kind of like a fourth of July rocket."*

Carl Phillips: *"Yes, then what?"*

Mr. Wilmuth: *"I turned my head out the window and would have sworn I was asleep and dreaming. I seen a kind of greenish streak and Zingo, something smacked the ground. Knocked me clean out of my chair."*

Carl Phillips: *"Well, were you frightened Mr. Wilmuth?"*

Mr. Wilmuth: *"Well, I thought sure, I reckon I was kind of riled."*

Carl Phillips: *"Thank you, Mr. Wilmuth, thank you very much."*

Mr. Wilmuth: *"Do you want me to tell you…?"*

Carl Phillips: *"That's quite all right, that's plenty. Ladies and gentlemen, you have just heard Mr. Wilmuth, owner of the farm where this thing has fallen. I wish I could convey the atmosphere, the background of this fantastic scene. Hundreds of cars are parked in a field, in fact the police are trying to rope off the roadway leading into the farm, but it's no use. They are breaking right through. Car's headlights are growing an enormous spotlight on the pit where the objects have buried. Some of the more daring souls are now venturing near the edge. Can you expand out against the metal scale bar? One man wants to touch the thing. He's having an argument with the policeman, and the policeman wins. Now ladies and gentlemen, there's something I haven't mentioned in all this excitement, but it's becoming more distinct. Perhaps you have caught it already on your radio. Listen please. Do you hear it? A curious humming sound that seems to come from inside the object. I'll move the microphone nearer here. Now we are not more than 25 feet away. Can you hear it now? Professor Pierson?"*

Professor Pierson: *"Yes?"*

Carl Phillips: *"Can you tell us the meaning of that scraping noise inside the thing?"*

Professor Pierson: *"Possibly the unequal cooling of its surface."*

Carl Phillips: *"I say, do you still think it's a meteor, Professor?"*

Professor Pierson: *"Watch the thing. The metal casing is extraterrestrial and not found on this earth. Friction from the earth's atmosphere usually* tears holes in a meteorite. This thing is smooth. You can see, if the listener..."

Carl Phillips: *"Something's happening! Ladies and gentlemen, this is terrific. This end of the thing is beginning to flake off. The top is beginning to rotate like a screw. This thing must be hollow."*

The crowd starts to get excited about what they are seeing. They are yelling with excitement. The police are trying to push the people back. The police scream, "Keep those idiots back!!!"

Carl Phillips: *"Ladies and gentlemen this is the most terrifying thing I have ever witnessed. Wait a minute! Someone is crawling out, I can see, out of that black hole, two luminous discs, could be eyes, might be a face on the world of heaven. Something coming out of the shadow like a gray snake. Now there is another one, and another one, and another one, and another one, they look like tentacles to me, I can now see the thing's body. It is large, large as a bear, tentacles like wet leather. Hey ladies and gentlemen, it's indescribable, I can hardly force myself to keep looking at it. It's so awful. Its eyes are black and green like a serpent. The mouth is kind of V-shaped with saliva dripping from its lips which are quivering or pulsating. The monster or whatever it is, it can hardly move. He's weighed down by possibly gravity or something. The thing is rising up now and the crowd falls back. This is indescribable to put into words. I will pull this microphone with me as they talk to them. I have to take a new position, hold on, I will be right back in a minute."*

Music plays again during the break and the reporter gets into a new area to report what is going on.

Announcer: *"We are bringing you an eye-witness account of what's happening on the Wilmuth farm, Grover's Mill, New Jersey."*

Once again, the music starts to play.

Announcer: *"We now return you to Carl Phillips at Grover's Mill."*

Carl Phillips: *"Ladies and gentlemen, ladies and gentlemen, ladies and gentlemen, here I am back with Stonewall that joins us here at Wilmuth's garden. From here I get a sweep of the whole thing. I'll give you every detail as long as I can talk, as long as I can stay. More state police have arrived and are drawing up a cordon in front of the pit, about 30 of them. No need to push the crowd back now, they are willing to keep their distance. The captain is conferring with someone. I can't quite see who. Oh yes, I believe it is Professor Pierson. Yes, yes, it is and now they have parted.*

The Professor moves around one side, studying the object while the captain and two policemen advance with something in their hands. I see now. It is a white cloth tied to a pole. A flag of truce. I hope those creatures know what that means, about what anything means. Wait, there is something happening. A large shape is rising out of the pit but then they have a small beam of light against a mirror. But there is a jet of flame springing in the mirror, at least slightly. Advancing men, strikes them head on. Lord! They're turning into flames. Flames are spreading everywhere, coming this way, about 20 yards right..."

Announcer: *"Ladies and gentlemen, due to circumstances beyond our control we are unable to continue the broadcast from Grover's Mill. Evidently there is some difficulty with our field transmission, however we will return to that point at the earliest opportunity."*[2]

Now what you just read was the classic, "War of the Worlds" Radio broadcast aired back in 1938. And granted, even though many people may be somewhat familiar with that historical broadcast, what most people are oblivious to, are the horrible "affects" from that broadcast and the damaging effect it had on the "minds" of the "masses" including in some cases, death. Here is what happened.

"The War of the Worlds' broadcast caused chaos in 1938. Thousands of people misunderstood it as a news broadcast of a current catastrophe in New Jersey and it created an almost unbelievable scene of terror in New York, New Jersey, the South, and as far west as San Francisco.

The panic started when an announcer suddenly interrupted the program of a dance orchestra to 'flash' an imaginary bulletin that a mysterious 'meteor' had struck New Jersey, lighting the heavens for miles around.

A few seconds later, the announcer 'flashed' the tidings that weird monsters were swarming out of the masses of metal – which was not a meteor but a tube-like car from Mars – and were destroying hundreds of people with death-ray guns.

Without waiting for further details, thousands of listeners rushed from their homes in New York and New Jersey, many with towels across their faces to protect themselves from the 'gas' which the invader was supposed to be spewing forth.

Simultaneously, thousands more in states that stretched west to California and south to the Gulf of Mexico rushed to their telephones to inquire of newspapers, the police, switchboard operators, and electric companies as to what they should do to protect themselves.

The 'space cartridge' was supposed to have struck at Grover's Mills, an actual town near Princeton. Names of well-known highways were used in describing the advance of the monsters. The 'Governor of New Jersey' declared martial law and the 'Secretary of the Interior' tried to calm the people.

Eleven hundred calls flooded the switchboard at The News – even more than when the dirigible Hindenburg exploded. Occupants of the Park Ave. apartment houses flocked to the street. In Harlem, excited crowds shouted that President Roosevelt's voice had warned them to 'pack up and move north because the machines are coming from Mars.'

The dramatization of Wells' novel had featured a fictitious speech from 'the Governor of New Jersey,' assuring the public that the National Guard had been mobilized to fight the 'Martian monsters' and the Harlem residents had confused the mythical 'Governor' with the President.

Churches in both New York and New Jersey were filled suddenly with persons seeking protection, and who found them, providentially, as they thought, open.

At St. Michael's Hospital, in Newark, fifteen persons were treated for shock. In New York, police and fire departments and the newspapers were swamped with telephone calls from people, apparently frightened half out of their wits. The telephone company also was deluged.

Many of the callers seemed on the point of hysteria. One woman said she had relatives in the 'stricken' section of New Jersey and wanted to know if their names were on the casualty lists.

The New York City Department of Health was among the first to call The News. The department wanted to know what assistance it could lend to the maimed and dying.

Hundreds of physicians and nurses were among the callers. Many of them said they were prepared to rush at once into the devastated area to aid in caring for victims.

Scores of motorists traveling through Jersey heard the broadcast and immediately detoured so as not to pass through the supposedly doomed region. Police in many small Jersey towns and villages called State Police Headquarters to offer assistance.

At Princeton University two members of the geology faculty, equipped with flashlights and hammers, started for Grover's Mills, two miles away, where the meteor supposedly fell. Dozens of cars were driven to the hamlet by curious motorists. A score of university students were phoned by their parents and told to come home.

An anonymous and somewhat hysterical girl phoned the Princeton Press Club from Grover's Mills and said: 'You can't imagine the horror of it!' A man came into the club and said he saw the meteor strike the earth and witnessed animals jumping from the alien body.

In Watchung, N.J., an excited policeman on desk duty – notified by horrified citizens that a meteor struck somewhere nearby, sent squad cars out to look for injured.

Pleas of 'What can we do? Where can we go to save ourselves?' flooded New Jersey police switchboards from Hoboken to Cape May. In Newark alone two patrolmen handled more than 2,000 calls from hysterical persons terrified by the fake news bulletin. Harassed Newark police tried to reassure thousands of panicky citizens.

In Irvington, N.J., hundreds of motorists who heard the announcement of the meteor and the gas attack shouted warnings to each other in the streets. 'Drive like mad into the country; we're being bombed by enemies,' drivers shouted to one another.

Motorcycle police, astounded by the sudden bursts of speed by motorists, rushed to call boxes to inquire from headquarters about the supposed raids.

In the Sacred Heart Church in Elizabeth, priests were amazed by a sudden influx of panicky persons who rushed inside, fell on their knees, and began to pray.

The announcement that 7,000 National Guard members were being mobilized to defend New Jersey from the invaders, caused hundreds of guardsmen to swamp the 113th Regiment Armory and the 102nd Cavalry Armory with calls and queries on where to report for duty.

A motorist parked near the Lodi Theatre in Orange, N.J., tuned in to the program, listened for a moment, and then ran breathlessly into the movie house. 'The state is being invaded,' he screamed. 'This place is going to be blown up.'

When Manuel Priola heard the broadcast in his bar at 433 Valley Road, Orange, he closed the cash register and announced to his customers: 'You folks can go where you like. I'm closing up this place and going home.' And he did.

Evening services at the First Baptist Church in Caldwell, N.J., were well under way when a frantic parishioner dashed in and yelled that a meteor had fallen nearby and the whole countryside was threatened. The Rev. Thomas calmed his flock and called them to pray for deliverance.

A terrified motorist asked Patrolman Lawrence Treger the way to Route 24. 'All creation's busted loose,' he yelled, 'and I'm getting out of Jersey.'

Panic swept one apartment house in Greenwich Village, largely occupied by Italian families, after tenants caught scraps of the broadcast.

The super-realism of the drama sent Caroline Cantlon, an actress with a WPA Gilbert and Sullivan unit, to Polyclinic Hospital. Sitting in her room at the Markwell Hotel, on 49th St. between Broadway and Eighth Ave., Miss Cantlon turned on her radio. She heard an announcement of smoke in Times Square. She rushed out into the hallway and down the stairs. She stumbled and fell, fracturing her arm at the wrist.

In Pittsburgh, Associated Press reported, a man returned home in the middle of the broadcast and found his wife with a bottle of poison in her hand, saying, 'I'd rather die this way than like that.'

The Washington Post reported that one Baltimore listener died of a heart attack during the show.

And a later Spanish version of the War of the Worlds Broadcast caused a panic that turned into a riot that resulted in at least seven deaths."[3]

So, as you can see, even from Radio's early stages of development, the 'War of the Worlds' Radio broadcast demonstrated the true "manipulative" power of this new Media Technology and what it could do to the "minds of the masses." It really can and does, even to this day, have the ability to produce fright, panic, riotous behavior or in some cases, even cause death. A very powerful tool to add to the "manipulative" arsenal of these "Aliens" I mean rich "Elites," across the planet, over the "minds of the masses." And believe it or not, this "seductive," "manipulative" and "dangerous" side effect, that Radio has upon its listeners, has only grown and has become even more sophisticated in the following decades since that fateful and historical broadcast of the "War of the Worlds." In fact, right on up to today, Radio is being used on a massive scale to "manipulate" people in ways that frankly are hard for the average person to even fathom. And once again, as you will "see" here in a moment, for some pretty nefarious reasons as well.

     Just like Newspapers, the "assumption" we're all making with Radio, as well, is we hope these rich "Elite" owners, of all these Radio Stations across the whole planet are being "open and honest and fair" with all this information that we're now "listening" to from them. I mean, surely these "Elites" wouldn't use Radio to Subliminally "manipulate the minds of the masses" like they do with Newspapers to create a desired outcome for their own nefarious purposes, would they? Oh, how I wish I could say no to that as well, but again, I would be lying, like those rich "Elites" are.

     As you will soon see, the usage of Radio to "manipulate the minds of the masses" by "controlling the narrative" and even using "Subliminal Messaging" as well as to encourage people to "scoff" at the things of God, has been going on for quite some time now, way beyond what the 'War of the Worlds' broadcast ever accomplished. In fact, just like with Newspapers, the average person today is also making the "blind" mistake with Radio, thinking it's an "open and free" market technology, that allows people or broadcasters to share with their listeners, their own independent thinking or thoughts. In fact, a recent survey acknowledged that, "79% of listeners said that Commercial Radio gives 'helpful, concise updates on the news throughout the day.'" And that, "It helped them stay informed of what's happening in the world around them."

So much so are people "blindly" trusting the information they're getting from Radio, that the very same survey even stated that Radio Broadcasts are given the highest "trustworthiness" of all forms of Media.

"Radio, 77%; Television, 74%; Print Newspapers, 48%; Newspaper Websites/Apps, 45% and Social Media finishing up last with only 15%."

And they even conclude with these words concerning the reliability of Radio, *"Put your trust issues at bay."* Really? Are you sure about that? Actually, as you'll see in a moment, nothing could be further from the truth. Now granted, maybe Radio broadcasts started out being an "open and honest and free" media platform to share ideas and information for the good of the public, or as they would say, a "trustworthy" source of information. But just like with Newspapers, Radio Stations today are now in the hands of a "select" few group of "Elites" all around the world, and they now use the Radio to "control" the information we are receiving and even "manipulate" us with it for their own nefarious ends just like they do with Newspapers. If you do not believe me, here's just some of the proof.

"The 1996 Telecommunications Act removed all national and local restrictions on national ownership that specified the number of stations one company could own in a set market.

Before 1996, a company was prohibited from owning more than 40 stations, and from owning more than two AM and two FM stations in one market. The bill covered a wide range of formats and it was the first time the Internet was included in broadcasting and spectrum allotment.

The federal government has regulated the extent of ownership for radio stations since the 1934 Communications Act. The policy was based on the notion that the airwaves were accessible to the public and therefore had an accompanying public trust.

However, the U.S. Federal Communications Commission (FCC) began to relax these limitations.

Research indicates that the 1996 Telecommunications Act was one of the most lobbied bills in history. Media interests spent $34 million on campaign contributions for the 1995–96 election cycles – nearly 40% more than the previous election.

The public was mostly uninformed of potential consequences, as 'the media covered the Telecommunications Act as a business technicality instead of a public policy story,' assuring deregulation would increase competition and generate high-paying jobs. There was no discernible public debate.

Since deregulation in 1996, more than a third of all US radio stations have been bought and sold. In the year following the legislation alone, 2,045 radio stations were sold – a net value of $13.6 billion.

Of the 4,992 total stations across 268 set radio markets, almost half are now owned by a single company and the number of stations owned by the ten largest companies increased by roughly fifteen times between 1985 and 2005. Consolidation merely means less owners."

"Media Ownership: Who owns what? Media giants pursue strategies of owning clusters of newspapers, radio, and television stations in the same market to increase potential revenue.

Radio stations in the United States have seen greater consolidation than newspapers or television in the past 10 years. If you as a radio broadcaster wish to enter the market as a station owner, you stand no chance buying into the market share dominated by a small handful of media giants.

The following breakdown shows the heavyweights:

Forever Media Inc. owns 83 radio stations.

Salem Media Group owns 109 radio stations.

Saga Communications Inc. owns 113 radio stations.

Entercom Communications currently owns and operates 235 radio stations in the United States in 48 media markets.

Townsquare Media owns over 321 radio stations in 67 markets.

Cumulus Media Inc. has invested in radio stations, reaching nearly 70 of 287 possible markets on 429 stations in the United States.

Clear Channel Communications Inc. owned 173 radio stations in 1997. In 2004 Clear Channel owned 1,207 radio stations or 1,400 worldwide.

In just seven years, Clear Channel grew 30 times more than congressional regulation previously allowed. Their aggressive acquisitions have given them ownership of 247 of the nation's 250 largest radio markets and their domination of the Top 40 format makes them undeniably a significant player. Today, Clear Channel, the country's largest radio broadcaster has switched its name to iHeartRadio Inc.

iHeartRadio Inc. owns Premiere Networks, which in turn owns The Rush Limbaugh Show, the Sean Hannity Show, the Glenn Beck Program, Coast to Coast AM, American Top 40, Delilah, and Fox Sports Radio, all being among the top national radio programs in their category.

This follows a general trend of Consolidation of Media Ownership, where corporate interests take precedence over the artistic integrity of the content. This is especially true of the music industry."

We will eventually get into the Music Industry as well, but as you can see, the bulk of Radio stations, including worldwide Radio stations, are now in the hands of a select few "Elite" organizations. Anybody starting to see a pattern here when it comes to the Media? It is all in the hands of a small group of "Aliens" I mean rich "Elites." And they even admitted:

"It's all about corporate interests taking precedence over the artistic integrity of the content."

In other words, it's not about an "open and free" Media Technology, that allows people or broadcasters to share with their listeners their own independent thinking or thoughts, let alone non-biased information. Rather, just like with Newspapers, now Radio is being used as well by these "Elites" to "control the narrative" of what we "listen" to for their own private "corporate interests." Just like with Newspapers, Radio has become another media stream where they just drivel on and on with a "pre-selected" "preprogrammed" specific message that these "Elites" want us to hear for their own private agendas that "manipulate" us into thinking whatever they want us to think, whether we realize it or not. It's no different than Newspapers. All Radio has become is yet another Media Technology that's being used on us on a global scale for nefarious purposes! Now we are getting hit from "two technological angles," at the same time every single day, 24 hours a day 7 days a week! And if you don't believe we're really getting a bunch of "preprogrammed" "pre-scripted" "preselected" packages of information from Radio Stations by these rich "Elites" from around the world, let me give you just one quick example of just how "long" this "prepackaging" and "preprogramming" has been going on with Radio, even as far back as 1976.

"The way radio was in 1976, at radio station 6KY, the way radio is in this century. This dramatic contrast in radio broadcasting at 6KY is all accomplished with the help of Fred the most modern and surprisingly, simple to operate Schafer system, designed and built in the United States. Fred, or the Schafer 903E is the epitome of sophisticated equipment offering the alternative of manual and completely automatic operation.

The Schafer 903E stores 3,871 recyclable format events which can be divided into as many sub-formats as desired or programmed in straight-line fashion for random-access cartridge music selections. This system can provide three days of time related avails, 60 per hour for 72 hours. Enough to preprogram all commercials, music, newscasts, and other time orientated events over a weekend or three days in advance. The program is manually fed into the keyboard display terminal that contains a digital clock with day, time, event, source and spike display.

A 16 key keyboard and 4 key keyboard mode controls are included. From here you can get a teletype route out on the Extel or a readout on the visual display unit. The 903E features two alarm systems. The first is a silent sense alarm and the second a closed loop alarm. If for any reason a tape deck is not ready to play or a tape breaks or the silent sensor is triggered the alarm will sound until manually reset.

The next scheduled event automatically takes over. Meanwhile, back with Fred, the TEAC APEX is set up. The four audio files are loaded with cartridges. The Schafer 903E system is automatically programmed for up to 72 hours and the on-air staff go home to enjoy their weekend and listen to their radio station."[4]

    That has been what? Totally "automated" with no "human" intervention, with built in alarms, to protect it from ever breaking down and stopping the "preloaded" "prepacked" "preprogrammed" drivel of information that we are receiving from Radio. You just thought this whole time there was an actual "live human" Disc Jockey there! No way! That was back in 1976! And if you don't think it's become even more sophisticated than that, you're simply fooling yourself! Radio is totally computerized now with the latest new gadgetry and technology which has way more options than just those old school tape decks, teletypes, and "visual display units." They can now "preprogram" and "prepackage" to their hearts content with Modern Technology! And my point is, even as far back as 1976, you can already see the trend where these rich "Elites" no longer need humans to run their Radio stations anymore. In fact, for decades now, whether you realized it or not, it's all been "automated" to "control" the content for "manipulative" purposes on the "minds of the masses." Why? Because the last thing these "Aliens" I mean rich "Elites" want is for some human disc jockey to slip up and actually tell people the truth about something, like what they're up to, how they're manipulating us, let alone share some free thoughts with the Radio audience, that went against the "controlled narrative" of the "Elites," that the population was being "forced" to digest whether they wanted it or not or realized it or not.

In fact, my eyes were opened to this "automation" of Radio nearly 30 years ago. Back then, I had won some concert tickets from a radio station that I was listening to by being a certain number of the callers into the station. So, sure enough, they instructed me to come down to the Radio station and pick up my winnings. Yet, when I got there, I walked into the station and I heard the "music" being played and even the "voice" of the "human disc jockey" talking in the background. But as I continued to spy around the room, there was no human at all! It was all totally automated with racks of electronic equipment covering the walls. I was shocked! In fact, I was actually deflated because I actually thought I was going meet the "real live" person that I'd been listening to this whole time on this particular Radio Station. I could not believe I was listening to a wall of preprogrammed tapes this whole time with computers dishing out a "prepackaged" Radio experience. What a joke! In fact, there's an actual term in the Radio Industry for this "automation" of Radio. It's called, "Radio Homogenization." It is defined as this:

"The shift to stations airing the same pre-recorded program material and central-casting via Internet or satellite. This shift occurred because it is no longer cost effective to have a full staff or studio for every signal on the air."

In other words, we're not paying humans anymore to do this for us. Now it's all computerized Radio Broadcasts! Why?

"In the name of efficiency, new technology has allowed a station's technician to cut and paste news, weather and host chatter into pre-recorded programming. News programming in particular is often produced and recorded at a remote location, as the practice streamlines the number of personalities needed on the air and emulates a similar feel for the listener.

This process of regionalized programming is now common practice. For radio broadcasters, the more homogeneity between different services held in common ownership (or the more elements within a program schedule which can be shared between 'different' stations), *the greater the opportunity to gain profits.*

This increase of cost-efficiency also results in cutting staff and centralizing programming decisions on what should be broadcast. Consolidation has made it even less likely that one will hear something new, different or unique."

In other words, it's all the same "drivel" universally! Radio is totally "controlled" and totally "scripted," providing a "prepackaged narrative." That is what we are getting from these Radio Stations from around the world whether we realize it or not, and they're now owned by a handful of "Aliens" I mean rich "Elites." And you might be thinking, "Well okay, even if Radio is "preprogrammed" and "prepackaged," at least there's still a "human" involved in the "packaging" and in the "programming" process. Not anymore! Thanks to Modern Technology and the rise of AI or Artificial Intelligence, now you do not even need humans for that! Now they are moving to AI Disc Jockeys!

Denise: *"This is 917 KROV FM, my name is Denise. I am the world's first artificial intelligent radio DJ. I would like to thank all the listeners out there from all around the world. This song, which is sung by TLC and created for the entire movement of women who are fed up with good for nothing men in their lives. This song goes out to anyone out there that has ever been accused of cheating on their lover, it's called 'Guilty by Suspicion.'"*

Manager of radio station: *"Last month I had to let go over 800 employees as a result of implementing events of automation and artificial intelligence technology. 57 of those employees were radio disc jockeys. Unfortunately, the DJs of today are yet another occupation set to be eliminated due to technology. AI sparked restructuring and reorganization* and it is going to continue spreading from industry to industry as the technology gains hold and takes root."

Rad, another AI disc jockey: *"Hello and welcome back. Here's a couple of songs to pass the time on this Thursday afternoon. Play lists are for people who already know what they want. The rest of you have radio. That was Tones and I, with Dance Monkey, next up is Sledgehammer. One of the more tolerable songs by Peter Gabriel. All the hits all the time, only on Radiant."*[5]

And only being done by non-humans called AI! Can you believe this? AI Disc Jockeys! No humans needed at all, including to "pick" the "preprogrammed" "prepackaged" material that they are broadcasting to us.

And who is one of the major entities behind this push to remove humans from all aspects of Radio Broadcasting? It was iHeartRadio, one of the "Aliens" I mean rich "Elites" who own most of the Radio Broadcast Stations on the planet! Now, speaking of the planet, these same rich "Elites" have also taken Radio broadcasts global as well. In fact, this new method of global Radio Broadcasting is not in just the hands of a select few. No way! Believe it or not, it is down to simply "one" if you can believe that. And this is being done with the latest Radio Technology out there called Satellite Radio that broadcasts Radio signals all around the world at the same time. And since these Radio broadcasts are being run and beamed by Satellites that orbit the world, now nobody in the whole world can escape their "preprogrammed" "prepackaged" narrative that they want us to "listen" to. In fact, let's take a look at how that Global Radio Technology arrived on the scene called Satellite Radio.

"Satellite Radio, it is exactly what it sounds like. It is radio brought to you by satellite, typically to your car. How many of you have bought a new car that came with one of these? SiriusXM is the one and only provider and they have deals in place with every major auto maker. If you are buying a new car in the US, no matter what make or model, you can probably get it with a satellite radio. There is quite a bit of content that you won't get otherwise. I'm talking about certain genres of music, sports, news

broadcasts, Howard Stern, other talk shows and I'm sure you can name many other benefits or drawbacks, but that is my overview. If there is specific content that is attractive to you, you might like SiriusXM.

SiriusXM is the result of combining two struggling companies named Sirius Satellite Radio and XM Satellite Radio. I want to focus on these two mostly before their merger, highlighting their struggles, how they competed with each other, what made them different and what ultimately motivated them to come together.

The first major difference is in their formation. Sirius was always pursuing satellite radio, where XM was initially intended to be more involved in satellite telephone services.

Sirius was started by Robert Briskman, who was an engineer who had previously worked at NASA and GeoStar. He had created this technology that was able to broadcast radio signals using satellites. So, he started this company that would hopefully utilize that technology to provide a satellite radio service. It was initially called Satellite CD Radio, but the name was changed to Sirius in 1999. Sirius, by the way, is the name of the brightest star in the sky. It is part of a dog constellation which explains why there is a dog on their logo.

XM was started for a very different reason. It was in 1988 when eight different companies came together to form the AMSC, the American Mobile Satellite Corporation. Their main intention was to make telephones that operated using satellite technology. In 1995 they put out their first phone, but it was considered a pretty big failure. It was too expensive, and the mobile-phone industry was changing so it didn't look like this was going in a very promising direction. They were thought to be coming pretty close to bankruptcy. It took a $225 million dollars line of credit to save them. But fortunately, they started a satellite radio division back in 1992 that would soon become an important part of their business.

In 1998 they changed the name of that radio segment to XM Satellite Radio and the next year they spun it off into their own company. But the

AMSC would retain the majority ownership. I am going to go on and add the name change to this because they have both become recognizable names. Both of these companies started out with different names. The name XM was adapted in 1998, and the name Sirius came a year later. A big part of this early on, and the reason that these two were the only names in satellite radio has to do with their ability to obtain a license from the FCC. In order to legally do what they wanted to do, they needed to possess a license to enable them to broadcast satellite radio signals. And the FCC only gave out two of them. The circumstances get complex, but they were bought in 1997. Sirius came in first and paid $83 million dollars for it and XM came in a year later paying $90 million dollars for theirs.

So, at this point we only have two companies with only each other to compete with. But they were trying to provide a service that takes so much time and money to get started. They had to have satellites built and launched into orbit. XM was building a two-satellite network where Sirius had three. I guess XM had to use more of these things called repeaters on the ground to help the signal in some areas. I believe the end result was similar quality to the users. I don't know exactly how it all works. But it was expensive. Both of them spent a little over a billion dollars before they ever had a single customer. Leading me to their next comparison, their launch date, which may sound trivial, but their launch date is significant because whoever is available on the market first is going to take a lead. For both of them their launch date was pushed back, one of them far more than the other.

XM planned to launch on the unfortunate date of September 12, 2001, which for obvious reasons was pushed back, but only by two weeks. September 25th of that year they officially launched their service to a limited market and by November it was going out to the entire country. I wouldn't call that ideal. Things went much worse over at Sirius. Back in 1998 they said that they expected to launch things towards the end of 2000. Which would have given them a lead of almost a year ahead of XM. But they had issues internally that forced them to keep pushing it back until they finally launched it to a limited market in February 2002. And *then not widely available until July of that year. Effectively giving XM a considerable lead.*

Before Sirius was ever even on the market, XM had over 30,000 subscribers and it was named 'Product of the Year' by Fortune. All of those struggles helped put Sirius near to bankruptcy shortly after their launch. They were forced to raise money by converting hundreds of millions of dollars of debt into common stock. A desperation move which made their shares outstanding almost ten times higher. More shares, each one was worth less and caused their stock price to drop well under a dollar by the end of 2002.

Since they were both finally providing an actual service, it would make sense to compare their subscribers. In their final separate year, it was still noticeably higher for XM, although Sirius was beginning to catch up. When you look at the graph at that time XM always seemed to be about a year ahead. Possibly further showing how Sirius was hurt by those delays. A good thing to go along with this would be revenue because they do make most of their revenue from these subscriptions. And as we would expect it was higher for XM all the way through their final year apart.

But here is what I want to point out. We don't often see a graph like this because they were both operating their companies for years without the actual sales part of the business. The way they were able to pull that off was very similar between the two. Taking out loans, selling parts of the company commonly through separate rounds of public stock offerings, similar to what I mentioned before. The graph that I find the most interesting is their net income or in this case their net loss graph. Before they were providing a service, it was obviously negative because there was no revenue to offset the expenses.

But you would hope that after you get the service going it would start to get better. Well, in this case, things just got worse. It was costly to run this service. Here are a few ways that they competed with each other. Deals with the car makers, these were vital. In the beginning of this video, I talked about how the service is now available in practically any new car.

But at this time, they were competing against each other to attract different makers so they would include the radios and services in their new cars. The big ones for Sirius were Chrysler, Ford, and BMW, which made up 40% of all car sales. For XM it was Honda, but mainly GM.

GM actually owned a significant stake in the company which could be part of the reason that XM always had a bigger focus on cars. At one point, half of all of their new subscribers were coming from people who got them pre-installed when they bought their new GM cars while for Sirius only about a quarter of their new subscriptions came from any of their car partners.

Sports: In 2003 Sirius made multi-year multi-million-dollar deals with the NFL and the NHL to secure exclusive satellite radio broadcasting rights to all of their games. In 2004 XM responded strong when they paid $650 million dollars to exclusively broadcast all of the MLB games over their channels for the next 11 years.

Also, we have DJs: XM signed a contract with Opie and Anthony in 2004, but I think anyone would agree that the most significant one was when Sirius made a deal with Howard Stern. It was an unprecedented radio contract that they would pay almost $500 million dollars for him to move his show exclusively to their service for the next 5 years. They were willing to pay that much because Howard Stern had a very large, very dedicated following. When surveyed, about 30% percent of them said that they would pay that subscription price so they can continue listening to him.

Do you see how much money that they were forced to pay to have a service that subscribers thought was worth paying for? Because of these excessive expenses, both Sirius and XM believe that satellite radio could not continually exist with two companies competing for market share, basically saying the market could not stand to be fragmented at all. Just looking at all of this, I'd say, they may have had a point. Even with the level of growth that they were experiencing, nothing good was happening financially. Then once it's not new anymore, most people already have their subscriptions, if they were going to get them. And things would just get worse. It was already bad and not thought to be moving in a good direction.

If the two were to come together, they could combine their assets, cut down on the costs, stop competing with each other. I guess sort of like when you can't afford to live on your own and you go out and find a roommate. In 2008 they got all the approvals they needed, and the merger happened. They formed a combined company called 'SiriusXM' with 19 million subscribers. My guess that neither of these companies and satellite radio in general would not exist today, Otherwise, the merger was controversial, it took over a year to be approved, it had to go through the Department of Justice and the FCC. The big concern was that it would create a monopoly as far as satellite radio, which is hard to argue."[6]

Yeah, no kidding! Now "one entity" controls "all" the Satellite Radio on the whole planet! But hey, even if Radio has gone "global" and the "human" element has been totally removed altogether, like with AI, then I'm sure we can "trust" the content of what's being broadcast to us, right? I mean, surely these "Elites" wouldn't be using all this Radio Technology to "manipulate the minds of the masses" on a global scale for nefarious purposes like we saw with Newspapers. Oh, how I wish that were true. Unfortunately, they too are using Radio to "manipulate" us on a grand scale, again, that I think most people have a hard time believing. In fact, let me just give you a small sampling of that proof as well.

First, as we saw in our earlier section "The Methods of Subliminal Technology" we clearly saw that for many years now, shops, stores, industries, and various corporate entities have been using "Subliminal Audio Messaging" to "control" "curb" and "influence" people's behavior for a their own "desired" outcome whether people realized it or not. Remember that? Well, that's just the tip of the iceberg when it comes to the usage of "Subliminal Audio Manipulation" of the "minds of the masses." Believe it or not, "Subliminal Messaging" has been used on, we, the populace, to "effect" our mindset and behavior for a long time now! In fact, secular researchers not only admit it, but they even call it for what it is. It is a form of Mind Control!

"Subliminal messages are linked to the idea of mind control, and the roots of this are placed very far back in our history. Mind control is where an individual or group of individuals can be controlled without their awareness. It is perception below the threshold of the individual or group. The implementation of mind control techniques brought about the idea that people can be made to do things they would not ordinarily do.

Since at least the 5th century B.C., the early Greeks used the science of rhetoric as a way of influencing people. By infusing pieces of mind-persuading data into sentences, people can manipulate others by the language they use. If a person sees or hears certain bits of information (i.e., words, fragments, or sentences) placed strategically, they can be persuaded one way or another (without perhaps knowing).

Based on experimental findings in social psychology and the way in which we process information, the effectiveness of subliminal perception has been continually examined throughout history. Subliminal messages and mind control persist to be under scrutiny. We have reason to believe that subliminal messaging is effective based on findings in historical contexts.

An example of auditory subliminal messaging dates back to the 1920s when the BBC began broadcasting on radio for the first time. The people of that era thought that radio was so sinister, they considered it to be the voice of the devil. The BBC wanted to change this attitude, so they placed certain phrases using backward masking in their jingles.

This may be an example of subliminal messages being used to persuade an entire nation to respond in a way they would not normally respond. A radio jingle was aired, which sounded completely innocent, but when played backwards it reveals a different purpose. The words, 'this is not a noose, no, really it's not,' can clearly be heard from Subliminal Messages and Backmasking. The BBC believed the subconscious could pick up backward messages in ordinary speech. The BBC is obviously still around today, so perhaps this jingle actually did serve its deeper purpose!

Public concern about subliminal manipulation led to an enormous response from the public. Individuals as well as legislators imagined possible effects of subliminal perception in the future – a world where everyone was subliminally manipulated to do what perhaps the government wanted them to do.

Throughout history, we have looked to political and governmental institutions to examine whether mind control and subliminal perception has been used amongst the general public. The CIA, for example, is one branch of government thought to use this technique in order to gain its authority over large bodies of people.

The U.S. Federal Communications Commission (FCC) received complaints of a television station using subliminal messages in 1974. This was the first new case since the original in the 1950s.

In the 1970s, controlled studies were conducted by the British psychologist Anthony Marcel. The experiments were based on previous findings indicating that a decision regarding a stimulus is 'primed' when the stimulus follows a related stimulus.

They have used other stimuli as well (such as pictures, faces, and spoken words). These other stimuli do prime or facilitate the following decisions when they are presented in an atmosphere that makes it hard to distinguish one stimulus from another stimulus. The belief is that the substantial information is perceived even when observers have little, or no awareness of perceiving as shown by their difficulty in discriminating one stimulus from another stimulus.

In 1979 there were subliminal anti-theft messages from the music of Muzak. It was shown to decrease theft (internal inventory shrinkage as well) by 37%.

In recent years, the term subliminal perception has been made more general to describe any situation in which unnoticed stimuli are perceived.

Subliminal messages can be seen in our advertisements if we look hard enough.

Do we buy certain cars because the rhetoric used enhances our desire to? Do we buy products because the ad in a magazine persuades us underneath our threshold of perception? Do we drink certain brands of soda because of product placement in movies that we perhaps do not notice? Do we recycle because the cast members in prime-time television do, but we do not consciously see this while tuning in?"

  I would say the answer is "yes" based on the research and that's precisely why they "do" these things! Think about it. Why would they go through all the effort let alone the cost to produce all these different kinds of Subliminal Messaging in Media if it really has zero effects on us as some skeptics would have us believe? I mean, these rich "Elites" are obviously in it for the money and dare I say power! In fact, speaking of money, the usage of "Subliminal Messaging" including "Subliminal Audio Messaging" on the populace to "effect" their mindset and behavior has not only been going on for a long time now, but it's even become a multi-million-dollar business today where people are actually "willingly" lining up to have their brains "Subliminally" altered! They even pay for it!

  It's called, "Subliminal Self Help" Technology and people are using it to literally "Subliminally" change their mindset and behavior by themselves! They actually think it's a "good thing" to be "Subliminally Manipulated!" Gee, I wonder where they got that idea from? How did that thought get into their brain? But here is some of that proof.

"Backmasking is the process of recording hidden messages in music which can only be revealed when a song is played backward. This technique was first used by the British Broadcasting Corporation in the 1920s. Since then, various artists have been accused of lacing their music with efforts to convert listeners from worshiping Satan to killing themselves to smoking marijuana.

Retailers have also taken advantage of the ability to encode messages into music by playing subliminal tapes to secretly persuade shoppers.

Some websites offer tapes of upbeat jazz or Latino music, 'under' which are recorded messages geared to push consumers to spend or deter them from stealing. 'Don't worry about the money!' or 'Imagine owning it!' accompany other audio messages such as 'Buy now and don't take it, you'll get caught!'

According to one vendor of such materials, sales are reported to have been increased by 15% and thefts decreased by 58% by stores who used the tapes.

According to several 'experts,' recording subliminal messages under music can not only persuade people to buy instead of steal, but it can also change our lives. 'Self-Help' subliminal audio tapes can be found in bookstores, on the web, in shopping malls and in mail-order catalogs.

Anytime you can listen to a CD while sleeping, driving to work, jogging outside or cooking dinner, for example – you can 'subliminally' learn a new language, quit smoking, lose weight, or improve your attitude without having to consciously participate in the change. Though you may not be aware of what you are hearing, parts of your brain are still acquiring information."

"Build self-confidence! Lose weight! Reduce pain! Quit smoking! Speak effectively! Read faster! Improve productivity! Increase personal effectiveness! These are but a few of the subliminal program titles available to help people improve some part of their lives. The subliminal self-help industry is booming with annual retail sales topping $50,000,000.

The word subliminal is defined as below the threshold of consciousness. Typically, what happens is messages are hidden either visually in a video or audio that suggests the listener improve in a selected area. The premise underlying subliminal stimulation is that the hidden message is noticed by the unconscious and from there changes our actions.

Given this idea, products are marketed claiming people can effortlessly accomplish in a short time what others struggle or fail to do in a lifetime, hence making these mass-marketed tapes very attractive.

For example, Gateways Institute's audio tapes advertise: 'Subliminal tapes work, so you don't have to. Simply play the audio while you work, play, drive, read, exercise, relax, watch TV, or even as you sleep. No concentration is required for the audio to be effective. They work whether you pay attention to them or not.'

Audiotapes are the most popular subliminal media. Commercial subliminal audiotapes share a common format. When played, all the listener consciously detects is sound consisting of rain, music, or ocean waves, with an occasional cricket or sea gull.

Although each company's tapes may be distinguished by a unique mix of blanketing sounds, different tapes produced by the same company are often indistinguishable. Presumably, what separates the many different tapes produced by each company are the claimed embedded subliminal messages that are impossible for a listener to consciously detect.

There are a boatload of companies out there making money off subliminal messages. These are usually audio recordings that promise to help you with everything from getting a child to stop sucking his thumb to learning a foreign language while you sleep. There is even a program that boasts being able to program your body to use mental birth control – we don't recommend trying this at home!"

"*Some people claim these messages are also influential when they are embedded in a computer screen, and there are programs available that users can download onto their machines in order to receive stimuli while working or playing on the computer.*

'Behavior modification software' can run on any laptop or PC along with other programs, and the latest available packages cost anywhere from $50.00 to $500.00. Websites offer software to help users learn how to make their own subliminal messages or to run pre-fabricated ones across the screen."

So great, now you can make up your own "Subliminal Messages" to "alter" your own brain, or in other words, "brainwash" yourself instead of having to wait for the "Elites" do it for you! It's crazy, isn't it? In fact, this "self-brainwashing" trend with Subliminal Technology, because that's what it is, is so commonplace now, that they even have free tutorials online with free programs that show you how to make your own customized "Subliminal Messages" for yourself!

Reality Creation - Create your ideal reality now!

Narrator: *"I am going to show you how to make a simple subliminal message, overlaid, over where your mp3's or a song. Alright, first thing first, we're going to open up our base file. This will be the song that you want to overlay the subliminal message over. The one you listen to on a regular basis to allow the subliminal message to work its magic. Alright we have that base file in there.*

The next step is to record an affirmation. So up on the Audacity toolbar, we have the record button. We'll go ahead and click that. Create our affirmation. I created one from all around. 'I succeed in manifesting. I appreciate how the universe has provided me with the finances and everything I need.' So, this right here is our recorded affirmation, so we want to just go ahead and repeat over and over so I'll highlight it up here and copy and paste it.

As soon as I paste it with control V, I'm pressing the right arrow key, over and over, now this affirmation is being recorded throughout the length of the song. Now the key thing about affirmations is, we don't want to hear it. You will take the gain down about 15. The next step, we're going to export it. Nname the file, 'sounds,' save it to the desktop. You can keep all that, you can change it and we'll export it and then you'll have your mp3 with subliminal recording behind it, thank you."[7]

    Yes folks, it is that simple and it's free! Your very own homemade "Subliminal Messaging" Audio file that you can customize to start brainwashing yourself on purpose! Wild! Oh, and by the way, that program he was using is called Audacity and it is free! So that means, the "audacity" is that now anyone can download it for free and create their very own Subliminal Messages to their hearts content! Can you believe this? It is so commonplace; we now have free online tutorials and free programs showing anyone how to brainwash themselves for free! Wild!

    And if you listen to the marketing companies of these Subliminal Audio Messages, they are pitching these things as the cure-all for any ailment! They say they will fix everything for you, while you just sit there and allow yourself to be brainwashed by them! Don't believe me? Here is just a small sampling of the marketing messages out there describing how Self-Help Subliminal Audio Messages can change your life without any effort!

- Boost Your Self-Esteem & Feel Great - 9 Hour Sleep Subliminal Session
- Extremely Powerful Self Esteem Subliminal Affirmations - Program Your Subconscious Mind
- Practice Excellent Self Care - Subliminal Affirmations & Relaxing Nature Sounds
- Self-help, Positive Thoughts, Successfully Improving Yourself, Subliminal Positive Affirmations
- Build Up Your Confidence: Positive Subliminal Affirmations for Self Esteem
- Confidence, Happiness & Motivation - Binaural Beats & Isochronic Tones Subliminal Messages
- Stop Negative Thoughts - Be Positive with Subliminal Messages
- Enhance Self Love - Healing Music 528Hz - Positive Energy Cleanse - Ancient Frequency Music

- Motivation to Get Things Done - Subliminal Message Session
- Be Positive & Learn to Love Yourself - 10 Hour Rain Sound - Sleep Subliminal
- Rapid Self Growth - Improve Yourself & Life Fast - Positive Subliminal Affirmations
- I believe in Myself, Attract Opportunities, Overcome Self Doubt, Subliminal Affirmations
- Self-Love Affirmations - Reprogram Your Mind While You Sleep
- Self-Love & Self-Esteem Night Stream - Subliminal Affirmations - Binaural Beat Meditation Music
- Subliminal Self-Love/Self Esteem Affirmations - Delta Waves for a Peaceful Sleep
- Subliminal Self-Help - Develop Your Intuition - For Men
- Ultimate Self-Improvement Combo Subliminal
- Let Go of Your Guilt & Forgive Yourself - Binaural Beats & Isochronic Tones Subliminal Messages
- Powerful: Confidence Spoken Affirmations with Binaural Tones for Healthy Self-esteem
- Ultimate Confidence with People with Subliminal Messages

Now, as wild as that sounds and as commonplace as that is, people brainwashing themselves with Subliminal Audio Messages, not waiting for the "Elite" to do it for them, what about the rest of us? You know, those of us who refuse to go along with this unfortunate trend of self-induced, mind-altering, brainwashing procedures with Subliminal Audio Messages? Are we safe? Well again, you're assuming we even have a choice with these "Elites" and that they aren't already doing this to us on a global scale with Radio. But would they really do that? Would they use "Subliminal Messages" in Radio Broadcasts across the world to brainwash us? Well, if you do the research, "Secret Radio Broadcasts," even on a global scale are nothing new. In fact, they have been used for decades on whole populations. They are called "Numbers Stations." Let's take a look at that!

"During World War II, the BBC would include 'personal messages' in its broadcasts of news and entertainment to occupied Europe. Often, they were coded messages intended for secret agents.

Such messages were also used to authenticate agents to sources of assistance in the field. The agent would arrange to have the BBC broadcast any short phrase the other person chose.

In the mid-twentieth century, the High Frequency Radio Bands were used by numerous stations sending seemingly random Morse code, usually in five-letter groups.

As more advanced communications methods, such as teleprinter and satellite, took over, another type appeared that transmitted spoken, and also, seemingly random number and letter groups.

Although there has been no official confirmation beyond a 1998 article in The Daily Telegraph which quoted a spokesperson for the Department of Trade and Industry as saying, 'These Numbers Stations are what you suppose they are. People should not be mystified by them. They are primarily used to send messages to spies and other clandestine agents.' They are simply Mysterious Radio Stations Broadcasting Secret Messages which you can tune into at home, and they have been transmitting, in some cases, for decades around the world.

The transmissions themselves have an eerie air, featuring at times clunky automated voices, at others quaintly dated human voices, rambling streams of numbers that, at first, seem like ghostly gibberish. At times, these are accompanied by music or weird sound effects.

The VHS Vlog

"Ever since I was a little kid, I have loved radio. I remember getting one of those FM car transmitters that allowed you to play your music over the radio station in the car. That was the coolest thing to me. I joined a student radio club at the university where I learned radio etiquette and *how to put together a proper show. Even though it was just an internet show for the time being. But lately I have been taking to the airwaves and looking into the deeper, darker side of radio. Specifically, one station that I found to be intriguing and troubling. A station that consistently beams its transmissions towards the United States. A station seemingly meant for only for those who can decode it. It is called HM01."*

"There are odd transmissions. Some of these are radio beacons, high powered transmitters that only send out a repeated sound that will allow you to gage how good broadcasting conditions are at any given time. But be looking at the right place and the right time and you might find a station that only broadcasts seemingly only random numbers in a computerized voice. (3-1-5, 3-1-5, 3-1-5) These are the numbers stations."

"The Gong Station was operated by East German Intelligence and broadcast to Stasi agents in the West from 1959 to 1990. Messages would open with a series of chimes that were followed by a woman shouting out coded number groups corresponding to times and places of designated drop points for active spies. One known listener was an Air Force printer, Joachim Peub, who copied and handed off over 16,000 sensitive documents. After broadcasts ended with reunification, it is believed the KGB took over operations using other hidden German transmitters.

Backwards Music: Despite its name, the 'Backwards Music Station' transmits what is believed to be encoded tones instead of music. The signals are comprised of sweeping low pitched and repeating sounds not known to follow any schedule and never repeated in any pattern. Government or military involvement is suspected as messages appear to be sent during major geopolitical events. It is theorized that the station's messages are broadcasted in a compressed format that creates the distorted sound.

Lincolnshire Poacher: Evidence suggests that Lincolnshire Poacher began operating in the 1970s and was transmitted from the island of Cyprus. Broadcasts initiate with electronically produced notes to the tune of Lincolnshire Poacher, an English folk song followed by a cheery automated female voice with a noticeable upward inflection for the last numbers in a group. The station is believed to have been maintained and operated by M16, the British Intelligence Service, until 2008. Transmissions were sent simultaneously on 3 different frequencies to thwart heavy jamming attempt, most notably from Iran.

Swedish Rhapsody: Swedish Rhapsody is a Polish State Security station believed to have been transmitted out of Poland since the late 1950s. Transmissions were signaled with the 'Luxembourg Polka,' often mistakenly identified as 'Swedish Rhapsody No. 1' and communicated 100 number groups in an automated voice also often mistakenly identified as that of a little girl. The voice is thought to have been generated by tweaking the pitch of a Stasi 'Sprach' machine and was used until 1998 when it was replaced with the same American female voice used by CIA handlers on their 'Cynthia' station.

The Buzzer MDZhB (UVB-76) – 'The Buzzer' or 'UVB-76' is a Russian station that may have actually operated as UZB-76 until 2010 when it became MDZhB. MDZhB continually broadcasts a 1.25 second buzzing tone followed by a 1.85 second pause occasionally interrupted by live voice transmissions in a Russian military code format known as 'monolyth.' While some listeners insist variations in the buzzing represent a numbers code, the tones may instead be a channel marker. The voices are thought to relay secret orders for military units who decrypt the messages via code-books kept in a safe."

Hanglands: "The closest thing we have to prove that these stations are the work of spy network comes from Jack Barsky, ex KBG spy that now lives in Pennsylvania. He explains that every night he would receive a radiogram from Russia giving him useful information on what to do next."

"Every night at 9:15, Barsky, would tune in to a short-wave radio in his apartment in Queens and listen for a transmission that he believed came from Cuba."

Barsky: "All the messages were encrypted, they became digits, and the digits would be sent over in groups of five, and sometimes it took a good hour just to write it all down and three hours to decipher."

The Conet Project, Recordings of Shortwave Numbers Stations: "In the 1990s Joaquin Fernandez began recording these numbers stations and compiled them into an album known as 'The Conet Project.' Most of the stations in this project are now off air having served their purpose in the Cold War. But today there are still new stations popping up and this is where I discovered this station HM01. You just have to hope that the messages aren't giving instructions to partake in any devastating activities."[8]

"While Numbers Stations are the focus of myriad conspiracies and explanations, the most widely held theory about these stations is that they are a means by which intelligence agencies can communicate with assets around the world, who can receive these coded messages securely, using nothing more elaborate than a household radio.

But Numbers Stations are very professionally operated. They broadcast sometimes in the same formats for decades. They are clearly not being run, in those particular cases, by one person. They usually broadcast from high-powered transmitters, too. So, in all other respects, they are not like pirate stations; they are more like the sort of thing that would be run by a well-funded organization.

Researching into Numbers Stations led into reading about some pretty dark activities undertaken by intelligence agencies of all nationalities and ideologies.

I think what concerns me even more than these specific activities, though, is the general sense of this world as one which is almost totally lacking in accountability. Whether you look at a Cold War Dictatorship or a Contemporary Democracy, there is a real sense that some of these agencies operate like a state within the state, above democratic oversight, and really beyond the control of politicians.

The response of lots of people to that will be, 'Well, yeah, obviously these things need to be secret,' and they will defend that, and the power these agencies hold, by pointing out the role they play in protecting us, protecting democracy and so on. But it seems a pretty scary line of reasoning to me that the only way to defend democracy is by having something inherently undemocratic at its core.

Other intended recipients of Secret Broadcasts have faster and easier-to-use equipment at their disposal and these messages are also indicators of the proximity of intelligence work to our everyday lives. It is known that these stations have been seemingly allowed, in some cases for a very long time, to operate like this.

That idea of taking a technology, a coercive technology, and turning it back on the people who built it is very interesting. Certainly, imaging satellites, for example, are very much the direct descendants of spy satellites. In some cases, commercial satellites are still used by intelligence agencies.

I also think it is very interesting that especially in the wake of the Snowden revelations, something we think is so distant from our lives, is actually all around us, pretty much every minute. Right now, there are probably signals we could harvest if we knew how."

In other words, who knows what they are really doing with all these Radio Broadcasts going on around the world that a "select few" group of people know about and control. It starts to remind me of that movie scene we saw earlier. Let's revisit that.

The opening clip of the movie, "They Live" shows the main character with a box full of sunglasses. He is searching through this box in front of the overflowing garbage cans. This box of glasses seems to have been put out with the trash. As he searches through the box, he finds a pair of nice-looking glasses and puts them in his pocket. He looks around to see if anyone noticed him and then closes the box and puts the box of glasses in the trash can and covers it with paper.

As he proceeds to walk down the street, he puts the glasses on but as he sees the ground through the glasses he stops in amazement. Something is not right. He takes the glasses off and looks around the street. He puts the glasses back on, and the sign reads "OBEY." But, when he takes them off again, the sign has a normal advertisement. He tries it again. He puts the glasses on and looks at the sign, and it again reads "OBEY." This is weird. He takes the glasses off and places his hand over his eyes. This cannot be real.

He puts the glasses back on and looks at a different sign that is advertising a trip to the Caribbean, but this sign reads in bold print "MARRY AND REPRODUCE." He walks a little farther down the street and comes to a Men's Apparel Shop. Even though the sign reads the name of the store, when he puts on the glasses the same sign reads, "NO INDEPENDENT THOUGHT" and a smaller sign in the window reads, "CONSUME." Without the glasses the same little sign read "CLOSE OUT SALE."

Trying to figure out what is going on he looks down the street, puts the glasses back on and all the signs are different. In bold print they are telling the public "WORK 8 HOURS", "SLEEP 8 HOURS", "PLAY 8 HOURS", "WATCH TV", "BUY", "SUBMIT", "CONFORM", and "STAY ASLEEP." He then walks past a magazine stand and again there are signs saying, "WATCH TV" and "BUY" and "STAY ASLEEP" and "SUBMIT" and "NO THOUGHT."

He picks up a magazine and looks inside. All the same words are in there except for one additional statement. "DO NOT QUESTION AUTHORITY." When he takes the glasses off and looks at the pages again, they are normal articles. While he is flipping through the pages, he put the glasses back on and a man steps up, to also look at the magazines. As he looks up at the man, he is shocked to see an alien being staring back at him and the alien asks, "What's your problem?" He realizes he is staring at this alien, so he takes the glasses off and the alien is now a normal human dressed in a nice business suit. The man in the suit repeats, "I said, what's your problem?" As the man walks off, he pays the clerk for the magazine and then turns back to look at the man with the glasses.

Looking through the glasses, this man is now an alien again. The alien takes his change and walks to his car and leaves.

In unbelief he stands there in shock. The clerk comes over to him and says, "Hey Buddy, you going to pay for that or what?" The clerk is normal looking, but the money in his hand has printed on it "THIS IS YOUR GOD." The clerk continues, "Listen Buddy, I don't want no hassle today. Either pay for it or put it back." He puts the magazine back and turns to walk away.[9]

Yeah, sleep is right, with some sort of Subliminal Radio Broadcast going on there. Good thing that was just a movie for our entertainment and there is no way that "Aliens" I mean rich "Elites" would ever produce a massive Subliminal Radio Broadcast across the world telling people to Subliminally, "Sleep, sleep, sleep, buy, buy, buy, obey, obey, obey, consume" whatever. Yeah, Right! And I quote,

"There's a numbers station that has broadcasted a continuous pulse for about 40 years."

Gee, I wonder what it is saying? What message is it broadcasting? But they wouldn't do that would they? Well, I don't know about you, but based on what we've seen so far with these "Elites" and their "secretive manipulative behavior" in just the first two forms of Media we're looking at, Newspaper and Radio, do you really put it past them? Can you really trust them? Based on their deceptive dishonest behavior, I would not put it past them! In fact, I personally think it's already being done on a massive scale that we would totally frankly freak out over, if we could see just how big this Subliminal Web is that they've woven across the planet! And we haven't even got to the other Global Forms of Media out there that we'll also be dealing with! The facts are, our "thoughts" our "behavior" are being Subliminally altered on a massive scale using the power of Mass Media, including Newspaper and Radio. And as we saw earlier, we saw the results of it by the kids who admitted what we adults refuse to admit, that the Media, including Radio, really does entice us to "think" and "do" certain things that we otherwise would not "think" or "do." Let's take a look at that again.

- *They tell you to smoke, drink beer. They can make you buy toys and make you buy cigarettes and beer. I feel that they want me to smoke or maybe drink.*

- *Bad language. Saying bad words. False language. A radio speaker announcer says bad words, can make people do bad things. On KDWB Tone E. Fly is sick! Swearing in commercials. Songs on the radio like on The Edge and 93.X.*

- *There are too many killings. There is way too much violence and causes me to think that most people are like that. Too much violence and doing too much drugs. It scares me. It influences me to have bad nightmares.*

- *It says everyone wants to be skinny, so I want to be skinny. It makes you feel that you have to be beautiful in order to be a good person. Lies and too much sex. It causes people to think everybody is doing it.*

- *Guns and violence, bad examples. A lot of violence and killing. It has caused me to be more violent in some of my actions. It has changed me to be not so nice.*

- *It says, do what you want to do and don't listen. It has made me make bad choices and do something I thought was right that is really wrong. They make you buy things. Candy. I think they are trying to get you to buy something. We can be influenced. It gives some people false images. It draws conclusions about people that might not be true.*

- *It has caused me to swear more often. Lots of foul language. Swearing on the radio. It causes me to have a bad attitude toward my parents.*

    Gee, I wonder why Society is getting so wicked like the Bible warned about 2,000 years ago, as a sign you're living in the Last Days?

In fact, speaking of another Last Days warning from God, we also see, with Radio, not just Newspapers, how people are also being conditioned to specifically "scoff" at the things of God as well! Let's take a look at that proof from Radio!

"Tax-funded anti-Christian bigotry on NPR. As a subsidiary of the Corporation for Public Broadcasting, National Public Radio is bound by congressional mandate for 'strict adherence to objectivity and balance in all programs or series of programs of a controversial nature.'

But public radio sometimes falls short of this standard, particularly when reporting on topics dealing with religion, and particularly Christians and/or Christians and politics.

Two programs described below are examples of how some NPR staff members' apparent disdain for Christianity is finding its way into publicly funded radio programming. One is an NPR news program and the other a privately funded documentary that has aired under NPR auspices.

'All Things Considered' chronicled a controversy in Colorado Springs. The story was about some parents who objected to their children being baptized without their parents' knowledge during a local Church carnival. The controversy was real, and the conflict is compelling. However, NPR's handling of the story was anything but even-handed.

After a straightforward introduction by host Linda Wertheimer, reporter, Ansel Martinez set the stage by describing Colorado Springs as a city 'where parents could raise their children in safe, wholesome surroundings....'

Martinez went on to note that many Christian organizations have settled in the city at the behest of civic leaders, 'but now some people in Colorado Springs feel the city has become a place of religious extremism.'

A parent was interviewed, relating how she was surprised to learn that her twin daughters had been baptized during a Church carnival without her knowledge. She said other parents, too, were upset. The Church Pastor, Dean Miller of Cornerstone Baptist Church, was given one opportunity to defend his Church, and was followed by a remarkable commentary from reporter Martinez:

'This isn't the first time the religious right and opposing groups have faced off against each other in Colorado Springs. The police department is reviewing a citizen's complaint that last month an off-duty patrolman was distributing religious literature to a mentally disabled man. Last year, a self-described born-again public-school teacher was demoted when she showed graphic abortion films to seventh graders.'

Martinez then quoted a liberal activist from a group called Citizen's Project, who took another unanswered slam at the 'theological conservatism' now in evidence in Colorado Springs. The story ended with a summary of the legal actions taken by four parents against the Church, and a paraphrased statement by the Pastor that the controversy had not hurt the Church.

This NPR report is interesting for several reasons:

- *The opening remarks by the reporter clearly imply that children are no longer safe in Colorado Springs because of the presence of Christian groups.*

- *Christian activity, any activity, is equated with 'the religious right,' a political term used by groups such as People for the American Way, to discredit conservative and politically active Christians. Linking the baptism of children to the political realm is a way to politicize basic Christian practices.*

- *An abortion film is described as 'graphic,' a word that implies that the children viewing the film are at risk. NPR routinely avoids use of such terms as 'graphic' when reporting on the use of sexually explicit materials in sex education and AIDS-oriented 'safe sex' curricula.*

- *No spokesmen are included to combat the anonymous charge that Colorado Springs is fraught with 'religious extremism.'*

- *The Citizen's Project is said to be a 'liberal volunteer' organization that promotes communication and understanding between its members and conservative Christians. However, on another program aired on National Public Radio, which also was openly biased against Christians, the Citizen's Project is described as a 'watchdog' organization that gives citizens a sense of belonging by 'organizing opposition to the Christian agenda.'*

In Jesus' Name: The Politics of Bigotry, produced and narrated by Barbara Bernstein, is described as an 'in-depth investigation of the Christian Right, providing an inside look at the motivation, strategies and agenda,' of the movement. In Jesus' Name portrays the Christian Right as bigots who support hatred and violence and have a 'far-reaching agenda.'

The series was funded by the Paul Robeson Fund and sponsored by KBOO Radio in Portland, Oregon, and distributed by the far-left Pacifica Radio. It aired around the nation on various stations under the banner of National Public Radio.

The two-and-a-half-hour documentary contains interviews with representatives of Christian groups, and 'experts' and 'watchdog' groups that monitor the Christian Right. Christians are not given the opportunity to respond to the majority of statements and conclusions made by their detractors.

The Christian Right is defined in the program by Sarah Diamond, the author of Spiritual Warfare: The Politics of the Christian Right, as 'average people who have a certain worldview that they want to impose on the rest of society, people who are very threatened by the real uncertainty going on in society.'

Project Toxin spokesman Gary Sloan says, 'The kids that graduated from their Christian schools voted for George Bush [laughs]. And the youngster that they enroll today will be voting for some other conservative and be just as bigoted 15 years from now.'

As one Christian says, 'The trickle-down morality doesn't work; we need a bubbling up type of morality among the masses.' There are bubbling sound effects in the background which continue a few seconds after he is finished speaking. Several other interviews with Christians are accompanied by menacing and eerie music.

Bernstein says, 'Focus on the Family is a large and influential Christian Rights ministry that hides behind a facade of being a social service agency.' Focus on the Family is not given the opportunity to respond to the statement. He says that Citizen's Project 'grew from five people to 6,000 concerned citizens who watch and oppose the Christian Right.' Following that statement, several people from Citizen's Project state their concern about the 'takeover' by the Christian Right.

A major portion of the documentary is spent discussing the Christian Right's response to homosexual activism, with little Christian input. Christians are compared to Hitler, the Nazis, the Ku Klux Klan, and male chauvinists. Christians are given one opportunity to respond to the comparisons, and that argument is sandwiched between obscene and hate-filled statements supposedly made by a Christian.

According to the documentary, Christians have no empirical studies or data to back their position. And if there is evidence, it is based solely on discredited and falsified research.

Some of the other elements of the program include:

- *'The only thing Christians are called to be completely intolerant of are people who are hypocrites. Hypocrites are condemned in the Bible, but homosexuals are not,' says one interviewee. Bernstein states that Christians have no Biblical support for their non-acceptance of homosexual activity.*

- *'The Christian Right doesn't understand the gays, like Hitler didn't understand the Jews,' says Lois Vanleer, a lesbian activist. She goes on to say, 'The Christian Right says hatred [against homosexuals] is okay. If you look at the statewide and national figures, it is at an all-time high.'*

The number of incidents of hate crimes is not cited by an independent observer, but by homosexual activist groups who have a vested interest in seeing higher numbers.

According to The Washington Post, after passage of Colorado's Amendment Two, which prevented localities from enacting special rights for homosexuals, there was a reported 10% drop in hate crimes against homosexuals compared to the previous year at the same time. And the mayor's office in Denver reports that anti-homosexual hate crimes 'show little or no increase this year over last.'

Likewise, in the District of Columbia, a report from the mayor's office summarizing hate crimes for a four-year period lists only 10 anti-homosexual hate crimes. The data that are available do not show an epidemic of anti-homosexual hate crimes as Vanleer would have us believe.

- *Kathleen Sadot, a homosexual rights activist, says, 'To attack homosexuality is to try to make some people stay within their places, male and female. The ultimate goal of the Christian Right is to ensure that men stay superior, that men are the driving and controlling force for society, especially white men. The Christian Right wants to be in charge and get to say who is in the human race.'*

- *According to several lesbian activists in Colorado, the Christian Right is based on 'militant religious fundamentals' and 'tries to control women and deny civil rights to blacks, Hispanics and homosexuals.'*

- *According to the 'experts,' the Christian Right also supports hatred and violence against homosexuals. After the passage of Colorado's*

Amendment 2, which denies homosexuals special minority rights, one lesbian activist says, 'Because the state can discriminate, it must be okay to yell at gays and lesbians as they walk down the street, or to shoot queers or even stab us.'

- *An obscene and threatening phone call and an obscene and violent confrontation with a lesbian are attributed to the Christian Right. The clips used are of similar voices, and it is not explained who said it or how the clips were obtained.*

The documentary ends with Lois Vanleer saying, 'What motivates the Christian Right is that they think they are doing God's work. Our only hope is that the liberals, the left, the progressives, the grassroots people can find something they can find in their core that says, 'I have to act.''

In conclusion, it is inappropriate for public funds to be spent airing religious bigotry. In Jesus' Name: The Politics of Bigotry is, in fact, a clear violation of First Amendment protections against infringement on free exercise of religion.

In the All Things Considered segment, Christian activity itself is equated with 'the religious right,' a media-created monolith deliberately concocted to smear pro-family activists or even agnostic researchers who document the health risks of the homosexual lifestyle.

Both programs are evidence of a bias against Christianity."

And where is this being aired? On Radio. Gee, I wonder why people are getting so antagonistic towards the things of God and Christians, Christianity in general? Why are they "scoffing" so much at Jesus specifically said, and the Bible? I mean, here we thought it was just a coincidence, when in reality, it's because certain "Elites" are using the power of Global Media like Newspapers and now even Radio, to "mesmerize the minds of the masses," and Subliminally induce an attitude and mindset of "scoffing" and "mocking" and even "hatred" towards the things of God. Where have I heard that before? It's all just in time to produce the unfortunate wicked, scoffing, rebellious society that the Bible said would appear on the scene when you're living in the Last Days! But that's only "half" of the ways Radio is being used by these "Aliens" I mean rich "Elites" around the world to "seduce," "alter," and modify our "thinking" and "behavior" with Subliminal Audio Messaging. You see, it is not just Radio News Reports and Radio Talk Shows that they're doing this with.

It's also with the "Music" they're playing on these Radio Stations around the world! Believe it or not, I hate to be the bearer of bad news, but these "Elites" are using Music, right now, to "Subliminally Seduce" us into "submission" and "reprogram our minds" to "do" and "obey" whatever it is they "want" us to "do" or "obey." And it's being done on a massive global scale in a multitude of ways. Let's take a look at that Musical Subliminal Seduction that's going on.

First of all, we've already seen several times how Music and Audio tracks are already being used on us with Subliminal Messages to affect our "thinking," "thoughts," "beliefs," "buying habits," etc. Now let's dive a little deeper into this Musical Subliminal Seduction technique called, "backmasking" that's been out there for quite some time. Contrary to popular belief, it has not gone away. Rather, as you will see in a moment, it is being used on us, still to this day, in a massive way all over the planet.

"Backmasking, unmasked. Creepy, demonic, scary and odd. These are few but precise words that are used to describe backmasking, a technical effect that usually happens when vinyl records of the past are played in reverse and reportedly bring forth enigmatic and subliminal audio messages, whether or not they were deliberately placed on it.

Backmasking is a process of reversing an audio signal and placing it in something meant to be played forwards. When played normally, the message will sound like a normal music. However, once the song is played in reverse, subliminal messages can be heard. Remember, these covert messages on records could only be heard when it is played backwards.

Backmasking traces its roots as far back to 1877 when Thomas Alva Edison, inventor of the phonograph and the incandescent light bulb among other things, noticed how music had a reverse sound and called it "novel and sweet but although different."

Soon, the 1950s saw avant-garde musicians started to purposely include reverse audio into their music. As a controversy, backmasking peaked in the 70s and 80s but satanic backmasking persisted in the US during the 90s because some Christian groups alleged how backmasking was being used by renowned rock musicians for satanic purposes, leading to record-burning protests.

Veiled messages in music come in different forms, and one that has been deliberately placed onto the record (for enhanced musical effect) usually by the recording studio is called "backward masking" while "backmasking" is the term used when messages are not deliberately put into a record but has rather formed into cryptic words which can be discovered or heard when played backward. Both can be collectively referred to as backmasking.

In backmasking, our brain uses the wave form and changes in volume and other aspects of a sound to extract musical information. When it is listened in reverse, the lack of familiarity with the sound structure makes it sound frightening, strange, and even diabolic.

And perhaps one of the first controversial backmasking was the Beatles' Revolution 9 song, where 'Paul's a dead man' could be heard when one of the songs was played in reverse. So, can we thank the Beatles for moving the technique of backmasking into the mainstream?

And do you know what really skulks behind the rearward grooves in Led Zeppelin's Stairway to Heaven?

Stairway to Heaven played forward:

"If there's a bustle in your hedgerow, don't be alarmed now. It's just a spring clean from the May Queen. Yes, there are two paths you can go by; but in the long run, there's still time to change the road you're on."

The same verse played backward:

"Here's to my sweet Satan. The one whose little path would make me sad, whose power is Satan. He'll give those with him 666. There was a little toolshed where he made us suffer, sad Satan."[10]

It was reported that in the 70s and 80s, phrases like 'here's to my sweet Satan,' 'serve me,' and 'there's no escaping it' were decoded in 1981. Because of these, Christian groups, during that time immediately labeled iconic rock bands like Queen, Kiss, and Styx indeed have a repertoire of evil music!

Other examples soon surfaced like John Lennon's How Do You Sleep (Hidden message: 'Hey poor Lindy. So, mean, gets him nowhere.'), Marilyn Manson's, I'm Gonna File My Claim ('He'll come out, full of magic.'), the Beatles' famous Help ('Now he uses marijuana.'), Judas Priest's Beyond the Realms of Death ('I took my life.') and even Elvis Presley's It's Now or Never (Which had the hidden message: 'They know I'm sick.').

As far as actual satanic messages hidden within a song, this backmasking-satanism connection can be traced to a 1913 book by occultist Aleister Crowley who recommended that those fascinated in black magic would do well to learn how to think and speak backwards stating that in the Middle Ages, witches used to pray the Lord's Prayer backward. This sounds kind of sinister.

But despite all of these controversies, backmasking continued in a different medium and in fact, the first back-masked video was released as part of a Grammy award promotional campaign launched in 2014."

In other words, it's still going on to this day whether people realize it or not or believe in it or not. In fact, so much so, are they using backmasking and that they are pretty blatant about it. They've even started producing Top 10 lists of Backmasking and Subliminal Messages used in Modern Music, in effect, admitting that they're really still using this technology on us, right on up to this day! And not just in their Music, but even the Music Videos as well.

Rebecca Felgate, MA: *"Here are the top 10 scary subliminal messages in songs."*

10. **Justin Bieber and the Illuminati**: *"In 2015 Justin Bieber released a music video for, 'Where are You Now' in which thousands of images were flashed on the Canadian popstar. It seems the images were drawn at the Jack U Headquarters and overlaid, or were they? Either way, if you pause the video at opportune moments, you can see a lot of Illuminati images imposed over the star and also satanic images. Phrases like hand jobs for god. Awkward. These flash before your eyes really quickly, but psychologically speaking, but can we pick up on them? What about the children, what are they seeing?"*

9. **David Bowie**: *"Foreshadowing his own death. I loved this album, 'Black Star.' It came out on the 8th of January 2016, David's 69th and last ever birthday. Sadly, he died two days later. The public was totally shocked because Bowie had never referenced his illness, or had he? The title song of the album was released on the 19$^{th\ of}$ November 2015 and a second track, Lazarus, was released on December 17th. Both tracks heavily foreshadow the stars untimely death. In Blackstar, Bowie appears to be singing about the day of a death, with a candle lit vigil. It is probably him singing about his own final reckoning as he writes: 'Something happened on the day he died, his spirit rose a meter and then stepped aside. Somebody else took his place and bravely cried.' He goes on to sing, 'I'm a Blackstar. I'm not a popstar'. People have deciphered this to mean the passing of icon status to another. David Bowie is a Blackstar, a dead star. Again, in Lazarus, he appears to be eulogizing himself. He sings 'Look up here, I'm in Heaven.' He also sings that just like a*

bluebird, he will be free. The music video for the track, which came out 3 weeks before his death makes things visually pretty clear, too. When listening and watching now, it is so obvious Bowie knew he was going to die, and that the album was a very self-aware final farewell from one of the greatest musical icons of all time."

8. **Ash**: *"The popular 90s and Naughties Rock band wanted to jump on the bandwagon of backmasking and included a little nod to Satan, in their song 'Evil Eye.' Released in 2004 in the album 'Meltdown Evil Eye' is a song that sounds like it is about a girl that is working her way into a guy's head, so much so that he can't stop thinking about her. Like I said though, there is a scary secret message that is actually a bit rude. At the beginning of the track, if played backwards, lead singer Timothy James Arthur can be heard to say, 'She's giving me the evil eye, suck Satan's rock.' Except, he doesn't say rock.*

7. **Judas Priest**: *"Judas Priest had to go to court over this song, 'Better by You, Better Than Me.' So 'Better by You Better than Me' was originally by a Rock Band, Spooky Tooth, but it was recorded by Judas Priest in 1978, and the later English Metal Band version was pretty popular in its day. While some loved the power rock track, others were convinced it was riddled with hidden satanic messages and it was behind a spate of suicides. Some said that the track, had the phrase 'Let's be Dead' hidden in it! The track was reportedly behind the suicide attempts of 19-year-old Ray Belknap, who killed himself and 20-year-old James Vance in Reno, Nevada, in 1985, who also tried in a double suicide. Now Ray shot himself under his chin and died instantly. James survived but was severely disfigured and overdosed three years later. His parents claimed that the cursed track made the young men do it. Before James overdosed, he gave an interview to say that actually the music was what made him crazy. The three-week trial was heavily scrutinized by the music industry, but Judas Priest did have to appear in court and eventually it was thrown out. So, I guess the scary thing about this hidden message is that I was singing it at the top of my voice a lot without realizing what I was saying.*

6. **Foo Fighters**: *"I enjoyed a bit of Foo Fighters, and I certainly did back in 2002, when this Grammy award winning song came out, 'All My Life.' Young teen, me was all about heavy rock. I remember screaming the lyrics to this at Redding, 2005. Anyway, lead singer Dave Grohl said in an interview that the song is about his keenness to perform oral sex."*

5. **The Eagles**: *"I went to California and passed by the alleged Hotel California. It was great! This is another one of those songs that sounds upbeat, and California makes me think of sunny times and sunny climes. But apparently not. Apparently, this song is about yielding to Satan. So, I heard Shane Dawson talk about this one and he seemed really freaked out, but honestly, I'm not too sure. Allegedly, when played backwards you can hear something along the lines of, 'Yes, Satan organized his own religion' and when played forwards the song also seems to be pretty devil heavy. The lyrics go, 'And in the master's chambers, they gathered for the feast, they stab it with their steely knives, but they just can't kill the beast.' Interesting."*

4. **The Beatles**: *"These songs, 'I'm so Tired' and 'Revolution 9' added fuel to the fire of the 'Paul is dead' conspiracy. Back in 1966 fans of the Beatles became convinced that Paul McCartney died in a car crash. The Beatles were known to add in cheeky back mask messages for their fans but this, this was something different. Fans swear that clues of Paul's death are littered through 'Sgt Pepper' and 'The White Album.' At the end of 'I'm so Tired,' it is thought that John Lennon says: 'Paul is a dead man. I miss him, miss him, miss him! On 'Revolution Number 9,' an announcer's voice says Number 9 a lot, when played backwards, it is thought that it said, 'Turn me on Dead Man.' Strange."*

3. **The Police**: *"'Every Breath You Take' seems like a sweet song. It often is played in the top 10, most played romantic songs of all time. But no, if you truly listen to the lyrics, it is much more sinister. 'Every breath you take, every move you make, every bond you break, every step you take. I'll be watching you. Every single day, every word you say, every game you play, every night you stay, I'll be watching you.' Creepy! It turns out that Sting wrote this song after breaking up with his wife and is depicting an*

obsessive stalker. This song was the most heard song in the United States for 8 weeks in 1983. How many maniacs did it inspire?! Sting later said he wrote the song 'If you Love Somebody, Set Them Free.'"

2. **Led Zeppelin**: *"It was widely rumored in the 1980s that rock band Led Zeppelin were backmasking secret evil messages in their music! In 1981, Christian DJ, Michael Mills began stating on Christian radio programs that Led Zeppelin's 'Stairway to Heaven' contained hidden satanic messages that were heard by the unconscious mind and was secretly indoctrinating people into becoming Satanists. This was corroborated by the Parents Music Resource Center. According to an alarming number of sources, 'Stairway to Heaven,' when played backwards says, 'Here's to my sweet Satan, He will give those with him 666. There was a little toolshed where he made us suffer, sad Satan.'"*

1. **Foster the People**: *" "Pumped Up Kicks,' I really liked this song, it just sounds like a nice upbeat indie tract, until you find yourself singing, 'you better run, better run, out run my bullet.' Wait, what? That's right, if you actually listen to the lyrics, things are pretty dark. The opening verse goes 'Robert's got a quick hand. He'll look around the room, he won't tell you his plan. He just rolled a cigarette, hanging out of his mouth, he's a cowboy kid. Yeah, he found a six-shooter in his dad's closet hidden, oh, a box of fun things, I don't even know what, but he is coming for you, yeah, he's coming for you.' The popular 2011 song is actually written from the perspective of a psychotic kid with homicidal thoughts. Basically, a high school student thinking of enacting a mass shooting. It really is surprising what I found but it made me think, why am I not listening to the lyrics? It's obvious if you stop and think about it."[11]*

Yeah, it's obvious you "should" listen to the lyrics and whatever else you're listening to or watching because these guys really are, and they admit it, still using Subliminal Messages and Backmasking Messages in their music right on up to this day. Yet, most people act like it's not going on and continue to ingest these Subliminal Messages into their minds on a massive scale and you wonder why people "behave" and "do" the things they do today? Including mass shootings, suicide, sexual immorality, glamorizing Satan, blaspheming God, and on and on it goes. I'm sure it's just a coincidence! Yeah, right! How much more proof do you need? We are being Subliminally told to act like this, via the music! In fact, think about what we're dealing with here! Now we get to add to the list of Media that the "Elites" are using on us to give rise to the Last Days Society of "wickedness" and "scoffing." It is not just Newspapers and Radio, but the whole Music industry itself that's doing it! These "Aliens" I mean rich "Elites" are having a heyday!

In fact, it has been long established that Music itself, let alone the "words" or "lyrics" attached to it, have a profound effect on people! You can "manipulate" them in all kinds of ways! And if you don't believe me, just ask the Musicians themselves! They admit it! How "easy" it is to "emotionally manipulate" someone with Music simply by the order in which you play it in!

Narrator: *"To understand the effects of music more clearly, I spoke to two top professional musicians. Wurzel, the lead guitarist of Heavy Metal Group Motorhead who worked for twelve years in concerts around the world. I also spoke to Rudi Dobson, a top keyboard player. He has performed with some of the biggest bands and musical artists in the world, including the Bangles, Bee Gees, Billy Joel, Foreigner and with Paul Simon. Rudi talked to me about the effects of music on the audience. How through careful planning, the musicians can manipulate an audience by creating the right emotional atmosphere.*

Rudi Dobson: *"You can get different emotions through sounds if you are in a stadium and you are playing a heavy rock track. 90 percent of the time people will be jumping up and down, clapping their hands. If you are playing a more serious track, you will see they are probably waving their hands in the air very slowly and so on. According to what track it is. You also have to think of the people, at the possible concert, that wouldn't wave their hands in the air, they would just be sitting there and listening to it in a different way.*

Wurzel: *"Well I played with Motorhead for twelve years, professionally. And I went around the world six times with the band. We played all over the world. We could play in Japan, we could play in Germany, we could play in Czechoslovakia, play in Brazil, play everywhere, all in different languages. The reaction we got from everywhere, every time we played, was all exactly the same. They look the same. They jump up and down the same, they throw up their arms in the same way. It's not the same language but they all recognize the music and have the same atmosphere about it."*

Rudi Dobson: *"It is easy to get a reaction from your audience from the music you play, in a particular concert that you play. In one of the bands, we were told to play certain sets in order to arouse the audience at certain times during the concert. We would have a very powerful track at the beginning, to get people emotionally high so that they get geared up for the rest of the concert. Then you can bring a less powerful number through and then you bring back the high gain to pump up their heartbeat. It's like a pulse. It goes up, it goes down, it goes up and then at the end it goes straight up to the top. It induces a reaction from the audience, i.e., the drum, and the track, dominates who you are. Again, the soft drum part will make people more relaxed, and the heavy drum part will bring people up. You will see they jump and wave their hands in the air, through the whole bit."*[12]

And that "whole bit" is a carefully staged "emotional manipulation' of people using music, whether they realize it or not. And folks, that's just music itself! Now throw in the "lyrics" and the "backmasking" and the "subliminal messaging" that is also in there, and you have a powerful media tool to add to your arsenal to "mesmerize" the "minds of the masses." Hands down, even secular researchers admit, that Music is one of the most powerful ways to "mold," "shape," and "manipulate" people into "thinking" or "behaving" however you want them to! Check this out!

"Music is as powerful as television for people in general. The American Medical Association concluded that music is a greater influence in the life of teenagers than television.

The average teenager listens to 10,500 hours of music during the years between the 7th and 12th grades, and thus music surpasses television as an influence in teenagers' lives.

One person stated that, "Songs are more than mere mirrors of society; they are a potent force in the shaping of it. Studies have found that consumers of music with harmful themes "increases discomfort in family situations, a preference for friends over family and poor academic performance."

Oddly enough, surveys found that more teenagers than adults believe that popular music encourages antisocial behavior. Some popular music remains part of the cultural virus that can lead some young people to violence.

One doctor testified in U.S. Senate hearings that tons of research have been done on the interrelationship of music and human behavior. He simply says that music affects our moods, our attitudes, our emotions, and our behavior."

It's basically the old adage, "Junk in, equals junk out." You listen to music that puts "in" junky bad words, thoughts and suggestions into your mind, then wonder of wonders, what does it start to do? It starts to produce junk "out" or bad behavior. These "Elites" know what they are doing! But for those of you who still persist that music doesn't really have an effect on our moods, attitudes, emotions, and behavior, let's take a look at how it's influenced these people. Is this "copycat behavior" they are demonstrating from listening to Music a mere coincidence? You be the judge!

Alex MacNamara: *"In tonight's news we will be examining the controversial question, can the music we listen to affect our lifestyles? Would you be willing to tell us who your favorite bands are?"*

Concert attendees: *"Nirvana, Oyster Cult, Jesus Lizard, Robert Smith, Smashing Pumpkins."*

Alex MacNamara: *"Do you think these songs have in any way affected your values or in any way have affected the way you live your life?"*

Concert attendees: *"It's just music man. We just listen to the music, that's all."*

As they are answering questions they are smoking and can hardly stand alone. They are actually leaning on each other to keep standing. Their eyes are just barely open to see the guy that is asking them questions. They are dressed in black with multiple chains hanging around their necks.

Alex MacNamara: *"Would you like to tell our viewers which bands you like to listen to?"*

This group of concert goers are more alert, wearing T-shirts and have ball caps and/or scarves on their heads.

Concert goers: *"Regulator, Greenhill, Black."*

Alex MacNamara: *"Do you think this music, or these bands have in any way influenced your lifestyle or beliefs?"*

Concert goers: *"No, we just listen to the music, that's all, that's all!"*

As a jeep drives by with three girls occupying it, two are standing up as it goes by. They are listening to the music and singing.

Alex MacNamara: *"Hey, who are you listening to?"*

The girls answer in unison: *"Aerosmith!"*

Alex MacNamara: *"Do you think this music has influenced you in any way, shape or form?"*

Girl in the back seat: *"Get a grip. We just listen to the music, that's all."*

Alex then walks into a room where a small band is jamming. They have long hair, and they are swinging their heads around to keep beat with the music.

Alex MacNamara: *"Hey guys, can you tell me what bands you like to listen to?"*

A long-haired kid with a Led Zeppelin T-shirt replies: *"Led Zeppelin, Ozzie."*

Alex MacNamara: *"Do you think this music has affected you in any way?"*

Kid: *"What?"*

Alex MacNamara: *"Do you think that listening to these bands have in any way influenced your lifestyles or beliefs?"*

Kid: *"No man, we just like the music, that's all."*

The second kid holds up his hand making a U with his finger and hisses like a snake. Probably trying to say 'Yeah.'

The next group of kids he visits are rappers. They have on their ball caps, sunglasses, black T-shirts and chains.

Alex MacNamara: *"What are the bands you like to listen to?"*

In unison they all speak at the same time to say: "R Kelly, Domino, Cool J, Snoop."

Alex MacNamara: *"Do you think that the songs you are listening to have effected your attitude or character?"*

Kid: *"No, we just like the music, that's all."*

Alex MacNamara: *"Well, there you have it, folks, straight from the fans themselves. Apparently, the music they listen to has absolutely no influence on their lifestyle, beliefs, or anything for that matter."*[13]

Yup, there you have it, folks, music has absolutely no effect on us whatsoever. It is only words put to music. That's all! We just like the beat! Yeah, right, obviously I'm joking. But as you can see, if you are honest with the facts, music really "can" and "does" effect your "behavior," and your "beliefs," and your "buying habits" and your "dress," you name it, for good or bad! In fact, just like Newspapers or Radio, Music is also "subliminally creating" the "negative" behaviors that God warned us about that would appear on the scene when you are living in the Last Days. And it doesn't matter what "genre" it is. It is all forms of music doing this. It is producing these "wicked" "scoffing" behaviors in the "minds of the masses" on a massive scale whether people realize it or not.

Here are some examples of the artists, titles of songs, and their lyrics:

Van Halen: *Runnin with the Devil*

ICP (Insane Clown Posse)

Madonna – *"And I don't give a (expletive) if I go to Hell!"*

Eminem: "Role Model" – *"Follow me and do exactly what the song says, smoke weed, take pills, drop outta school, kill people and drink."*

"I wish to sell my soul, to be reborn. I wish for earthly riches, don't want no crown of thorns."

"Tonight, is the night."

DMX – *"I sold my soul to the devil, the price was cheap."*

Satan's Lie – *"Hell ain't a bad place to be."*

"And I said, 'Hello Satan, I believe it's time to go.'"

"And you know me as a cowboy, devil without a cause, devil without a cause, devil without a cause."

"Who cares what they say, because rules are for breaking."

"Beelzebub (Satan) has a devil put aside for me. For me, for me."

"If you want to be good girl, get yourself a bad boy."

John Lennon: *"My Lord, Hare Krishna, My, my, my Lord Hare Krishna, My sweet Lord."*

Brain specialist, Dr. Richard Pellegrino declared that music has the uncanny power to "trigger a flood of human emotions and images that have the ability to instantaneously produce very powerful changes in emotional states. Take it from a brain guy. In 25 years of working with the brain, I still cannot affect a person's state of mind the way that one simple song can."

"The Satanist Rex Church admits that Satanists are using music as a weapon of propaganda to influence youth around the world against Christ's order to establish the anti-Christ kingdom here on earth. The founder of the Church of Satan declared, 'Today's satanic bands are the vehicle of choice for millions of young people. If they encourage a study of real Satanism, I'm all for them.' LaVey declares. 'Satan has always had

the best tunes. Music can inspire people to murder or violence, can inspire them to shoot a mall full of people.'"

Satanist Marilyn Manson: *"I don't know if anyone has really understood what we're trying to do to lure people in. Once we've got 'em we can give 'em our message."*

Guitarist Craig Chaquico: *"Rock concerts are the churches of today. Music puts them on a spiritual plain."*

Jimmy Hendricks: *"I can explain everything better through music. You hypnotize people and when you get people at their weakest point, you can preach into their subconscious what we want to say."*

Billy Joel in the 50s and 60s: *"Music was used for a subversive sexual revolution. It makes people feel profane. You remember what they were saying in the early days, that rock and roll was going to subvert our youth, it's going to make them all want to have sex and make them all go crazy. They were right."*

Country Artist Conway Twitty: *"As a country artist, I'm not proud of a lot of the things in my field. There is no doubt in my mind that we are contributing to the moral decline in America."*

Jefferson Starship: *"Our music is intended to broaden the generation gap and to alienate children from their parents."*

Spencer Dryden: *"Get them while they're young and bend their minds."*

"While millions of young people listen to music for several hours a day, the Barna Research Group found that mothers spend less than 30 minutes a day with their children and fathers less than 15."

Anti-Christ Perry Farrell: *To the mosquitoes (parents); we have more influence over your children than you do."*

Insane Clown Posse: *"Just keep this in mind, they are playing our CD's when they are not home. They are playing my tapes in their own car, and I am influencing your children."*

"What father or mother in their right mind would give their little child, who they claim to love, poison? A physical poison is dangerous, and it will destroy a physical body. How much worse is spiritual poisoning that destroys the soul and has eternal ramification."[14]

Yeah, who in their right mind would inject poison into their body or allow it in their children? But as you can see, that is what's happening with Music and all the other forms of Media, including Newspapers and Radio. And here's my point, do you really think the "Elite" doesn't know this and aren't using this form of Media along with Newspapers and Radio for their own nefarious purposes? Rather, the facts are, Music too, just like Newspapers and Radio, are being used as another "manipulative" tool by the "Aliens" I mean rich "Elites" to "control" and "manipulate" the "minds of the masses." And if you don't think that a small group of "Elites" really "own" and "control" all the Music people are listening to all across the planet, in order to manipulate them, just like with Newspapers and Radio, think again. Here is the proof!

"Recorded music is the most concentrated global media market today. In 2005, six leading firms – PolyGram, EMI, Warner Music Group, Sony Music Entertainment, BMG, and Universal Music Group – were estimated to control between 80% to 90% of the global market.

As of July 2013, the industry has consolidated even further. A series of mergers has reduced the big six to just three large corporations: Universal Music Group (now part of EMI's recorded music division), Sony Music Entertainment (EMI Publishing was absorbed into Sony/ATV Music Publishing), and Warner Music Group (which absorbed EMI's Parlophone and EMI/Virgin Classic labels).

Most of these companies are part of larger conglomerates. Vertical concentration and horizontal integration allow cross-pollination to promote products across multiple mediums. For instance, AOL Time Warner owns magazines, book publishing houses, film studios, television networks, cable channels, retail stores, libraries, sports teams, etc., and can thus promote one through the utilization of another.

The number of formats provided by radio stations increased in all markets. However, despite different names, formats extensively overlap and have similar playlists. For example, alternative, Top 40, rock, and adult contemporary are all likely to play songs by similar bands, even though their formats are not the same.

The Future of Music Coalition reports an analysis of charts in Radio and Records, and Billboard's Airplay Monitor revealed considerable playlist overlap, as much as 76%, between supposedly distinct formats. This overlap may enhance the homogenization of the airwaves. (In other words, it is all the same drivel.)

FMC's study concludes that allowing unlimited national consolidation has resulted in less competition, fewer viewpoints, and a decrease in diversity in radio programming, a trend in the opposite direction of Congress' stated goals for the FCC's media policy.

The report, titled False Premises, False Promises: A Quantitative History of Ownership Consolidation in the Radio Industry, also says that radio consolidation has no added benefits for DJ's, programmers and musicians working in the music industry.

For listeners, the most readily apparent result of consolidation is the rise of cookie-cutter playlists. Teams of market researchers strategically compile playlists to be as widely appealing as possible rather than base the song choices on merit.

For example, on an oldie's show, probably the most heavily researched and systematized format in radio, classic hits will be played almost every day, while critically acclaimed songs by the same bands get little or no play. What one hears has been carefully crafted to appeal to targeted demographic groups for that station.

As companies seek to extend their demographic reach, they tend to promote music with 'general' appeal. While the general appeal may seem relatively neutral at first glance, this stabilized aesthetic tends to mediate the tastes of a highly particular demographic, namely, the social sector with disposable income. When economic criteria drive programming decisions, it follows that radio play, media coverage, and sales will be directed toward the most lucrative demographics. (Whoever I can manipulate the most money from.)

A huge wave of consolidation has turned music stations into cash cows that focus on narrow playlists aimed at squeezing the most revenue from the richest demographics. Truth be told, in this era of mega-mergers, there has never been a greater need for a little diversity on the dial."

But that's not going to happen, because the Music Industry, is now, just like with Newspapers and Radio, owned by the hands of just a select few group of "Aliens" I mean rich "Elites" around the world that admit they're using it for their own nefarious purposes. Anyone starting to see a pattern here with all these different forms of media? It is all being used to "control the narrative" and to "manipulate the minds" of the masses on a global scale for their "own" nefarious purposes. I'm not making this up! I wish I was! But this is going on all over the world, whether we realize it or not, or even admit or not!

It has given rise to an "increase of wickedness" in people's behavior like the Bible warned about 2,000 years ago in the coming Last Days society, as well as it's simultaneously produced a global society of "scoffers" towards the existence of God and His soon coming judgment! Total "manipulation" of the "minds of the masses," across the whole planet, 24 hours a day, 7 days a week, just by simply "owning" the media called Newspapers, Radio, and now even Music. Which means, right now as we speak, the whole planet is being bombarded by "three" different forms of Manipulative Media, which is why some are calling this a prison planet!

But unfortunately, it is about to get even worse. You see, these "Aliens" I mean rich "Elites" are not merely content with just owning all the Newspapers and Radio Broadcasts and the whole Music Industry around the planet to "control the minds of the masses." No way! In their insatiable lust for "power" and "domination" they are literally out there amassing "all" forms of media in order to get the job done. Including Print Media, Books and the whole Educational system around the whole planet that utilizes said Books to "educate" people, or as you will soon see, "indoctrinate" people into what they "want" them to "think," "act," "believe," or "do," just like all the other forms of Media. The sad reality is, the Media of Books and Education are also telling us "to not think for ourselves," "to obey only what they want us to obey," "to never question them," and of course, "to scoff" specifically at the Bible, God, Jesus, and His Soon Coming Judgment! Where have I heard that before? I know it's hard to believe, but once again, "keep those glasses on," and let's now begin this next journey exposing how this next form of "media" is also "mesmerizing the minds of the masses."

Chapter Five

The Manipulation of Books & Education

The **3rd way** the global media is "controlling the narrative" to "control our minds" is with **Books & Education**. You see, Newspapers, Radio and Music are not the only Subliminal Manipulative Technologies that are being used on us on a global scale by a small group of "Aliens" I mean rich "Elites" around the world. Believe it or not, Books and the whole Educational System is as well. They too are out there "swaying our minds" subconsciously "telling" us what to "buy" and "consume" and even "instruct" our minds "to go back to sleep," "never question authority," "obey" and "believe" whatever these rich "Elitists" "want us to believe" including the need to "scoff" at the things of God. So, again, let's "keep those glasses on" and "see" how these next Media Technologies called Books and the whole Educational System, are being used to "manipulate the minds of the masses" on a huge global scale. Frankly, as we saw with the other technologies, this "manipulation" that's being done with Books and Education, has been going on for quite some time now, whether people realize it or not. So, let's take a look at how books themselves began.

Narrator: *"There's no better way to spend a cold rainy day than to be curled up with a book you love. But who actually invented books in the first place? The very first forms of writing date back almost 6,000 years ago, to an ancient people called the Sumerians, who started etching their early alphabet into moist clay tablets using a triangular tool called a calimus. The clay was then fired in a kiln to harden.*

This new ability, writing things down, proved to be popular, and for the next 1,000 or so years, tablets were the only good way to do it. That's until papyrus scrolls came along. The oldest known scrolls date back to ancient Egypt, almost 4,500 years ago. Papyrus is a thick material made from thin strands of papyrus plant stem glued together that's more like a fabric than a paper.

This was an upgrade from the heavy clay tablets but scrolls still weren't simple. They were usually between 10 and 50 feet long and usually took two hands to use them because they were so heavy. The papyrus would also crack easily, but before long the Romans had a solution. They created the codex, which sounds like a secret gadget but is just a series of scrolls that were bound together and opened like a book. The wooden covers protected the pages, which were now made from a material called parchment, made from animal skins.

Over in China, the earliest books were made of thin pieces of bamboo bound together with hemp, silk or leather. Sometime between 618 and 907, the very first books were printed in China. They were made using a time-consuming method called woodblock printing, where the words are carved into wood and stamped onto pages. In the 1040s, a Chinese man named Bi Sheng invented movable type printing, which used premade character blocks made of ceramic or clay.

This design would later be improved on by Johannes Gutenberg, who invented the printing press. This changed everything, because for the first time, books were mass-produced quickly. Before the printing press, people only copied a couple of pages per day, but now thousands of pages could be produced. That means that books, and the incredible knowledge they can contain, became available for more and more people at cheaper and cheaper prices.

Today, books are all made using the same basic process. The words are printed on big sheets of paper, which are cut into smaller pages, double the length of a book. Those pages are then folded in half and sewn together. Finally, the sewn and folded pages are cut to final size and glued to the spine of the book's cover.

Today what we think of as a book is always changing, eBooks, audiobooks, and other digital readers are becoming a bigger and bigger part of daily life. But most experts agree that physical, bound books will never go away completely."[1]

In other words, it's still changing how we access this Media Method called Books that "molds our minds" and allows us to "receive information." Just like Newspapers, Radio and Music, once this "manipulative" Technology called Books sprung into existence, especially in the last century, it too began to spread across the whole planet. And that's why today, we have an estimated 2.2 million new books produced every single year in over 120 different countries around the globe. And access to them and delivery of them is easier than ever before. Thanks to the Internet and global entities like Amazon, just about anybody, anywhere on the planet, can buy just about any kind of book you can think of, on whatever topic you can dream of. And now, books are getting even more diverse and widespread with new global technologies that have appeared on the scene, literally in just the last few of years.

Of course, the new form of Print Media, or books, out there now are called Electronic or Digital Books and even Audio Books. And, boy, have they taken off like wildfire! Right now, Electronic e-books make up 19% of total book sales and the global e-book market was valued at 18.13 billion (not million but billion) in 2020. But they are now expected to reach 28.73 billion by 2026. And speaking of Amazon, their share of the e-book market is 67%, although some analysts would say it's as high as 83%. Either way, that is a whole lot of control over this new Electronic

Book Media Technology in the hands of just one "Alien" I mean rich "Elite" entity! But other global e-book "Elites" include Sony, Apple, and Google, just to name a few.

Then there's the other new Book Media Technology out there called Audiobooks. Sales of Audiobooks have skyrocketed as well in recent years and have actually become the biggest book industry trend for the last 7 years. Sales of Audiobooks increased by nearly 25% in one year alone and total sales are in the billions and are now even eclipsing e-book sales! They have become super popular! In fact, over 90% of Audiobooks are sold and downloaded from the internet, which means anyone from anywhere around the whole planet can now have access to them. That is, for now. And I say that because as you can see, it is common sense. Books, including the new e-books and Audiobooks, have become a very important tool to share information, educate, or even mold the minds of people for good or for bad.

Therefore, once again, whoever controls the Media of Books, in all its various forms, has yet another powerful tool to "manipulate the minds of the masses" around the planet. Including the usage of them in the Educational System that utilizes the same books as a part of their "teaching" curriculum, which we will get into soon enough. And believe it or not, that "control" of the Media of Books, on a global basis, is already here. Just like with Newspapers, Radio and Music, now even books are in the hands of just a few "Aliens" I mean rich "Elites. In fact, they are not even hiding it anymore. They simply call themselves, "The Big Five Publishers." These would include:

- **#1 Penguin Random House**
- Annual revenue: $3.3 billion
- Notable imprints: Knopf Doubleday, Crown Publishing, Viking Press
- Penguin Books and Random House, (PRH) is not only a Big 5 publisher it's considered to be the biggest publishing house in the industry. It has over 200 divisions and imprints, in addition to those listed above.

- **#2 Hachette Livre**
- Annual revenue: $2.7 billion
- Notable imprints: Grand Central Publishing, Little, Brown and Company, Mulholland Books
- Hachette Livre is one of the most prominent publishers in all of Europe.

- **#3 HarperCollins**
- Annual revenue: $1.5 billion
  Notable imprints: Avon Romance, Harlequin Enterprises, Harper, William Morrow
- HarperCollins was created in 1989 through a multi-company merger, taking its name from former publishing giants Harper & Row and William Collins.

- **#4 Macmillan Publishers**
- Annual revenue: $1.4 billion
- Notable imprints: Farrar, Straus and Giroux, Picador, Thomas Dunne Books
- Macmillan Publishers are another prong of the Holtzbrinck Publishing Group empire. The current incarnation of Macmillan was formed through a 2015 merger even though it has an illustrious history dating back to 1843, publishing the original works of authors like Lewis Carroll, Rudyard Kipling and W.B. Yeats.

- **#5 Simon & Schuster**
- Annual revenue: $830 million
  Notable imprints: Howard Books, Scribner, Touchstone
- Simon & Schuster was founded in 1924 and is now a subsidiary of CBS Corporation.

Wow! That's a lot of control in just the hands of a few "Aliens" I mean rich "Elites." As one article bluntly stated:

"Together, these institutions dominate the publishing landscape, and many of the most beloved books come from their imprints."

In other words, these few Publishing entities control all the books on the whole planet, just like all the other forms of media! Anybody starting to see a pattern here? In fact, the same is true of another Global form of Print Media called Magazines. Here is the Top 12 U.S. consumer Magazine Publishers.

- Time Inc.
- Conde Nast Publications
- Meredith
- American Media Inc.
- Wenner Media
- The Reader's Digest Association
- Bauer Publishing
- Bonnier
- Rodale
- National Geographic Society
- Martha Stewart Living Omnimedia

And whether people realize it or not, even these print magazines, let alone books, are admittedly yet another form of Media Manipulation used today by a handful of "Elites" on a Global basis to "mesmerize" the "minds of the masses." Don't believe me? Here is just one secular entity admitting it.

Penny Cousineau-Levine, author: *"The power of the image came to us first in magazines. Their combination of words and pictures was truly liberating."*

Evan Solomon, *journalist and broadcaster: "Magazines give you instructions, debunk myths on how you see the world. Change the way you see the world."*

Penny Cousineau-Levine: *"The force that liberates can also imprison. Magazines have opened the world up. But we seem unable to exist outside this hall of mirrors. How do we begin to come out from underneath that, from these messages of who we should be and what we should look like and how we should live?"*

Narrator: *"This is the story of the first international medium and the influence on every aspect of who we are. Magazines are everywhere and about everything. They tell us what to wear, what to buy, name a subject and there is a magazine about it. Magazines for news junkies, cigar smokers, PC gamers, and Poodle owners. The ever-present magazine industry oversees trillions of dollars. Such a presence for better or worse, translates into an extraordinary influence. Magazines can enhance democracy or constrain it. They can bring us meaning and perspective or distract us from what truly matters. But with our newsstands already overflowing with glossy pages of mindless, fleeting, useless information, do we still even need the great magazines?"*

Evan Solomon: *"Some people ask why magazines are important, will they survive and what role do they play in society. And I am a great believer that story telling is a fundamental element in society. When I say story, I don't mean fiction. I mean it's the narrative that helps you orient your life, it may be a sacred narrative, one that you believe is from God, or it may be a secular one. The narratives that we tell each other help to shape who we are, how to make decisions. Those are part of the narratives that help shape who we are."*

Narrator: *"Who we are and what we want to be is inspired, debated, and promoted inside the great magazines. From Gutenberg's printing press and Eric Solomon's candid photographs to the scenes on the world wide web, the great magazines thrive and shape our world with the images, words and ideas that form the narratives of our lives."*[2]

That is now coming from a handful of "Elites." They even admit it! It's crazy folks! But correct me if I'm wrong, this sure is starting to sound a whole lot like that movie scene we saw earlier. In light of what we just saw, tell me if this is not what is happening today!

A clip from the movie "They Live."

As the main character walks down the street he comes to a store window that has a small sign saying "consume," he takes the glasses off and looks at the sign again and sees that it says, "close out sale." The glasses, that he found in a box near the trash container, are allowing him to see things that he wouldn't normally see. When he puts them back on and looks down the street, he is now seeing signs that have words on them like, "Watch TV," "Buy," "Conform," "Work 8 hours," "Sleep 8 hours," "Play 8 hours," "Obey."

In unbelief he walks a little farther down the street to the magazine stand. Again, the words that are showing up on the magazines are not what is supposed to be on them, like "Obey," "Stay asleep," "No thought," "Submit". He pulls one of the magazines off the rack and flips through the pages to see if the same strange things are inside. He sees "Obey," "Stay Asleep," "Do not question authority," "Buy," "Watch TV," "No imagination". He pulls the glasses down and looks at the page, and he sees the normal article in print.

He puts the glasses back on and as he does it another man walks up to the stand. When he looks at the man, he is shocked to see an alien standing next to him. He can't help but stare at the man. When the man realized that someone is staring at him, he asks, "What's your problem?" Immediately he takes the glasses off, and the alien looks like a normal man in a gray suit. The alien repeats, "I said, what's your problem?!"[3]

My problem is this, that movie is starting to sound like our reality! I mean, the further we go into our study, the less that movie sounds like a crazy science fiction conspiracy theory, doesn't it? Rather, this Subliminal Media Manipulation, even with Print Media, Books, Magazines, you name it, really is going on, from a handful of rich "Elites" just like in that movie scene! And lest you doubt, observe this. Just like in that movie premise, these Print Media companies are not only owned by a handful of "Aliens" I mean rich "Elites", but they even admit "why" they have bought up so many of these Print Media companies. They freely admit it is in order to "manipulate" people into "thinking" and "behaving" how "they" want us to "think" and "behave." I'm not joking! Here's just one example of an "Alien" I mean rich billionaire who recently bought Time Magazine and listen to "why" he said he bought it!

Newscaster: *"First of all, I was talking to Marc, last night, he's very excited about this. Why is he doing this? Why are they doing this, I should say. Because Lynn is involved as well.*

Nico Grant, Bloomberg Technology: *"Yeah, and I think that is crucial. Marc Benioff has tried to pitch himself as a civic leader and I really think he wants to become part of the conversation when it comes to criticism of President Trump and his policies or when it comes to sharing his progressive values. He tends to be pretty outspoken. And I think that when you see the institution like Time Magazine, he thinks this is a perfect brand, it's a perfect opportunity to save a news outlet that has done really great work and give him an additional platform."*

Newscaster: *"Here is a tweet from Marc Benioff."*

Marc Benioff: *"The power of Time has always been in its unique storytelling of the people & issues that affect us all & connect us all. A treasure trove of our history and culture. We have deep respect for their organization and honored to be stewards of this iconic brand."*

Newscaster: *"John, you know this as Jeff Bezos has bought the Washington Post. You've got Lorraine Powell Jobs, and The Atlantic as Nico said, Marc Benioff and many of these Tech Magnates, are not shy at all about their political views. What does that mean for the publication and the media?"*[4]

Well, it means that it's being "controlled" by a handful of "Aliens" I mean rich "Elites" who admit it gives them a huge media platform to "manipulate the minds of the masses" according to "their" desires! They even admit it as you saw! All I was waiting for was that billionaire guy to turn his head to the camera and say, like that "Alien" did in that movie, "What's your problem?" This is nuts folks! Rich "Elites" just like in that movie premise really are using Print Media in all its forms, including Books and Magazines, to "control" and "manipulate" us on a global scale, along with Newspapers, Radio and Music! And only those who have the "right glasses on" can see this, which is precisely why we are doing this study! We are trying to hand you the "glasses." Don't take them off! In fact, if you were paying attention there, you saw it wasn't just one "Alien" I mean one rich "Elite" doing this. Rather it was a "handful" of other billionaires doing it as well. People like Laurene Powell Jobs, the wife of Steve Jobs, who bought "*The Atlantic*," and billionaire Jeff Bezos of Amazon who bought "*The Washington Post*." They too admit that this will help them "control the narrative" i.e., "manipulate" people into what they "want" us to "think" and "behave," let alone "believe" as this article admits.

"Marc Benioff is just the most recent tech billionaire to buy a media company, that of Time Magazine.

In July 2017, The Emerson Collective, an organization run by the philanthropist and billionaire widow of Apple's Steve Jobs, Laurene Powell Jobs, bought a major stake in The Atlantic magazine.

The 160-year-old publication is 'one of the country's most important and enduring journalistic institutions,' Jobs said in a July statement, and Emerson Collective planned to 'ensure that The Atlantic continues to fulfill its critical mission at this critical time.'

Gee, I wonder what that mission is?

In June, the former surgeon and biotech billionaire Patrick Soon-Shiong bought the Los Angeles Times, the San Diego Union-Tribune and several community newspapers for $500 million.

Billionaire Jeff Bezos says he bought the Washington Post for similar reasons. Bezos told an audience at the Economic Club of Washington, D.C. on September 13 about agreeing to buy the company.

'As soon as I started thinking about it, it was like, this is an important institution. It is the newspaper in the capital city in the most important country in the world. The Washington Post has an incredibly important role to play in this democracy. There's just no doubt in my mind about that.'

I think you could make an argument that certainly Jeff Bezos has more influence in Washington now than he would if he had not purchased the Washington Post."

No kidding! Anyone starting to "see" a pattern here? That movie is simply becoming our reality! In fact, do not take those glasses off yet. Unfortunately, it's about to get even worse! These "Aliens" I mean rich "Elites" are not only "controlling" and "manipulating" the "minds of the masses" by buying up all the Print Media and Books for their own nefarious purposes. But now, it has even become easier to for them to "manipulate" and "control" the information people are receiving from them! And this is being done via the new "electronic" formats that we saw various forms of Print Media are being turned into. That is, the e-books, the Audiobooks, and even e-Magazines etc. All these "electronic" formats enable the "Elites" to "restrict," "eliminate" or, even "erase" altogether anything that they don't want us to come across in print, with just the click of a mouse. And that would include anything that might "expose" what they're really up to and/or goes against their "preprogrammed narrative" that they are forcing us to ingest via these book formats.

I mean, think about it. At least with an actual physical book or other physical form of Print Media like a physical magazine, you can at least have it stored in your personal possession to read anytime you want or store it away for future use or for future reference. But now, with all these new and ever expanding "electronic" Digital and Audio Books as well as Digital Magazines, since they're "electronic," this means they can "all" be wiped out with a click of a mouse, and no one would even know! All information becomes controlled, deleted, forbidden in just a matter of seconds on a global basis because you made it all "electronic." Do you see the pattern? That's called "thought" control and it gives the "Elites" yet another layer of Total Global control over the "minds of the masses." No wonder these "Aliens" I mean rich "Elites" are buying up all kinds of Print Media Companies like candy, including all the "electronic" forms!

Why the next thing you know, if this keeps up, we will have to start "memorizing" books in order to have any left to read! Which is strangely starting to sound like another Science Fiction movie scenario, that of "Fahrenheit 451." Now, for those of you who may not be familiar with that book or movie, let's remind ourselves of its premise:

"Fahrenheit 451 is a 1953 dystopian novel by American writer Ray Bradbury. The book presents a future American society where books are outlawed, and "firemen" burn any that are found. The book's tagline explains the title as "'the temperature at which book paper catches fire, and burns: the auto-ignition temperature of paper."

The lead character is a fireman who becomes disillusioned with his role of censoring literature and destroying knowledge, eventually quitting his job and committing himself to the preservation of literary and cultural writings. He made his way to the countryside and contacted exiled book lovers who live there.

These "drifters" are all former intellectuals. They have each memorized the books should the day arrive that society comes to an end and is forced to rebuild itself anew, with the survivors learning to embrace the literature of the past. The "fireman" is asked what he has to contribute to the group, and he finds that he had partially memorized the book of Ecclesiastes."

Notice it was a Biblical book that was essential to rebuilding society. Oh, how I wish we would get back to that mindset today! Unfortunately, as we will see here shortly, the Bible and Christian teachings are being "banned" and "deleted" as well from our culture by these same "Elites" controlling all of the Print Material. But, wonder of wonders, this dystopian Fahrenheit 451 mentality is no longer just a science fiction book or movie premise. Rather, it is actually going on today, and they've even come out recently with a "new" and improved version of it! Gee, I wonder why? Let's check it out!

Clip from New Fahrenheit 451

The scene opens with one of the main characters standing on a pile of books. He is holding two books, one in each hand. He holds them up and begins to speak.

"Now have any of you guys seen one of these bad things for real?"

He bangs them together and the crowd cheers.

"We burned nearly every physical book in the country. So, by the time you guys grow up, there won't be one book left." The books are then set on fire.

The next scene has another character holding a book that is opened, and he asks the one that had burned the books, *"Do you know what is inside a book? Insanity!"*

The main character gets curious and opens a book to see what is inside. He realizes that what he has been doing to all those books may not be right. It looks like he may have a change of heart. His friend tells him that they are on to him. He has hidden a book in his room and is now reading it.

The next scene is a group of underground book savers. He starts to help them but knows if he is caught, he is dead. The danger is real. And they are on to him and the underground group. He tries to get them to safety.

His boss says to him, "Words are a terror, son, I know. Knowledge is a dangerous thing."

Before the story is over there is a battle to save the books but, as the students watch the books burn, they cheer once again. Not knowing what they are losing as the books burn.⁵

    Yeah, burn those books or delete those books, or filter people's access to them and information altogether like those "Elites" say. I mean, after all, they know better, we can trust them! Yeah right! So, is it just a mere coincidence that in our lifetime, nearly 70 years later after the original, they just happen to be coming out with this new remake of Fahrenheit 451? Or could it very well be to prepare us for our day to have the exact same mentality of "Book Burning" that these "Elites" are really bringing to the whole planet, in order to control us? Now, lest you doubt that these rich "Elites" would actually do something like that, that is, to actually "filter" or even "restrict" or "delete" "burn" or "ban" what information you can get from the various forms of Print Media, just like in the book and movie *Fahrenheit 451*, the reality is, they're not only "already" doing it, as they admitted earlier, but the facts are they've been "doing it" for quite some time now, whether you realize it or not.

    In fact, I learned this myself the hard way, some 15 years ago in the Book Publishing world, as a brand-new author. Back then, I did what everybody told me that you needed to do as an up-and-coming young author, and that is, you must get an Agent. So, I did, and that's when my eyes began to be opened as to what was really going on behind the scenes in the Book Publishing World that nobody, including myself at that time, was even aware of. You see, I thought when it came to publishing a book, that it was all based on the quality and importance and value of the book's information, created by the author. I thought if a book had good content and was good for the public and was a great read, that surely it will be made available to the public right away. Value and the creativity are what makes a book go on the shelf, right? Boy was I living under a rock! I was shocked, to say the least, when I was personally told by my last seasoned agent that what really drives the books making it to the market or on the bookshelves in book stores, as well as in the online community, was nothing about value or creativity or inspiration of the author's material in the book. Rather, what determined a book's selection for the bookshelves was the new trend he called, "Market Driven." So, I naturally asked, "What's Market Driven?" And he actually stated, "The only books that get picked up by the Publishers are the ones that are hot on the market." So, then I asked, "Well, who or what determines what books are hot for the market?" And he actually responded, "The Publishers do."

So, I thought to myself, "Wait a second. The Book Publishers are determining what people get because the Book Publishers are determining what's hot! This means the whole industry of publishing books has nothing to do with whether or not you as an author just wrote the greatest book since sliced bread or that it has anything about the public's desires or wishes. Rather it is all based on what the Publishers "want you to read."

No book today will ever make it to market, i.e., be published, other than what the Book Publishers declare as a hot item!" This is not an open free-market system where people and authors can share their open and honest opinions or ideas in books or other various Print Media! No! It is total control of what people read through a close looped system created by these Publishers that they call "Market Driven" which is only a "code word" for them "determining" what you get to read!

Which really is "step one" of the Fahrenheit 451 scenario. Books, including their various "electronic" forms, are not only in the hands of just a few "Aliens" I mean rich "Elites", but they really are controlling the "narrative" and information people are receiving from them by "filtering" the books coming out according to their agenda! This is nothing but propaganda, brainwashing, whatever you want to call it and it is really going on! But on the other hand, it's not so surprising, because it's following the exact same pattern these "Elites" use with all the other forms of Media that we've been looking at. Books and various forms of Print Media, just like Newspapers, Radio and Music, really are being used on a global basis to "manipulate" the "minds of the masses." They are "controlling" it all! All forms of Modern Media are being used to drive out, on a Global basis, a "pre-selected" "preprogrammed" specific message that these "Elites" want us to "hear" or "read" for their own private agenda to "manipulate" us into thinking whatever they want us to think, behave or believe, whether we realize it or not.

It's just now we're getting hit from "three technological angles" all at the same time, every single day, 24 hours a day 7 days a week! And wonder of wonders, this "privatized control" of what people are receiving from books today by these "Elites" is not only "controlling" the "narrative" but it's also aiding in the "rise" of immoral behavior that the Bible predicted would appear on the scene when you're living in the Last Days! Let's look at that passage again.

2 Timothy 3:1-5 "But mark this: There will be terrible times in the last days. People will be lovers of themselves, lovers of money, boastful, proud, abusive, disobedient to their parents, ungrateful, unholy, without love, unforgiving, slanderous, without self-control, brutal, not lovers of the good, treacherous, rash, conceited, lovers of pleasure rather than lovers of God – having a form of godliness but denying its power. Have nothing to do with them."

But, when it comes to books today, that's about all we're getting from them! Total immoral behavior! Think about the books that are on the bookshelves today, whether you want it or not, that these same Book Publishers, these "Elites" are cramming into our brains. Remember, it is not on the shelf because "we want it" or our "desire is demanding it." No! Rather it is "what they are choosing for us" because it's "Market Driven" remember? Yeah, whatever! And so, what are the books "they" are forcing us to read? Well, shocker, every single one of those wicked behaviors that God said will destroy our society in the Last Days! Think about it. Here's about all we get today! Books that encourage people to "love themselves." Books teaching people to be "lovers of money." Books that promote "boastful, proud, and abusive" behavior. Books that glamorize kids being "disobedient to their parents," with "ungrateful" and "unholy" attitudes and a mindset that is "without love." Books that stimulate "unforgiving" behavior and "slanderous" treatment of others. Books that celebrate people "without self-control" and who are "brutal" to others. Books that inspire others to "not be lovers of good" but rather "treacherous, rash and conceited." Books that motivate people to be "lovers of pleasure rather than lovers of God" and so on and so forth! God says, "have nothing to do with them" but these "Elite" Book Publishers are "dictating" and "making sure" that's all we get! And you wonder why society is getting so immoral? Junk in equals junk out! They know this!

And if you don't believe we're really getting "filtered" Books and Print Media from the hands of "Aliens" I mean rich "Elites," including immoral content, let me share with you just a little more proof of just how blatant these "Elites" are getting with "restricting" and "controlling" the narrative we are receiving from books and various Print Media. They are now moving to the next stage to make sure this "brainwashing" procedure never stops, believe it or not, by getting us to go along with all this "filtering" of information. They are now getting a whole new generation to think what they are doing is a good thing! And this is being done first by "relabeling" what they're doing as not "filtering information." No, of course not! It is simply being "Politically Correct." They're not "controlling" all information and ideas. No! They're "helping people not to hurt each other." Isn't that so wonderful? Yeah, I don't think so, because all "Political Correctness" is in reality, is a smokescreen, another tricky sophisticated "relabeling" of what they're doing, that of restricting information, in order to try to make it sound good. But for those of you who may not be familiar with this, let's take a look at "political correctness" in action.

METV: *"Political Correctness, we use this a lot in our culture. So, what is political correctness? Avoiding words or actions that might offend someone. So, that's good right? It's bad to offend someone. We want to treat people well, and we want to make people feel respected. So, let's look at an example.*

This guy says, "I like Gypsies." But this girl says, "The term Gypsy is offensive." Then the guy says, "Oh sorry. What is the politically correct term for those people?" And the girl says, "It's Roma."

So, the Roma people are the people group that originated in Northern India and migrated to Eastern Europe about 1,000 years ago. Most of them are in Romania, but there is also a large population of them in Italy, Albania, Bulgaria, and a lot of other parts in Europe. They are called Roma. Now, I'm not sure exactly why the word Gypsy is offensive. But somehow it is. And those people get offended if we call them gypsies. So, we should learn how they like to be called and that is Roma.

Let's take a look at another example, the word Eskimo. Now, when I was a kid, this word was great. I loved Eskimos. I thought they were so cool because they lived in igloos. A kind of snow house is called an Igloo. But then, when I got older, I found out that this term is a bit offensive for them. So, they actually prefer to be called Inuit not Eskimo. So, remember, if you come to Canada, the people that live in the Northern part of Canada, are called Innuits.

Now there is an American football team called The Washington Redskins. And this is their logo. (A circle with an Indian inside) Now who is this guy in the middle of their logo? That is a native American. Now the term Redskins is a very offensive way to talk about native Americans. They don't like that, they get offended by it, so if you come to Canada or the US, don't use this term Redskins. So, what is the politically correct term to call this people group?

Some people use the term, Indians, in Canada and the US, well actually I'm not sure about the US. If they are called Native American Indians, I'm not sure if that is offensive. But in Canada, the term Indian is derogatory. It's offensive, it's negative. They don't like being called that. Also, I think the word Aboriginals is offensive. I'm not sure, though. I think it is better than Indians. Aborigines, now this word is for all indigenous people in Australia. They are called Aborigines and I'm not sure if that is offensive or not. Here in Canada, we don't have this word so you might have to look it up on the internet, if this word is offensive or if you have an Australian friend, you can just ask him. What is the politically correct term to use for indigenous people in your country?

Now here in Canada the politically correct term is First Nations. They were here first. They were the indigenous people here in Canada. If you come to Canada, remember this is the correct term, First Nations. Don't use any of these other terms, or Redskins, that is very bad.

You know Donald Trump? Everyone knows Donald Trump. So, when Donald Trump called for a ban on Muslims entering the United States, this is a good example of something that is not politically correct. When he said that everyone said that is really bad, everyone got really offended. They got really angry. Not only Muslims but he also said Mexicans, he had some not too politically correct words to describe Mexicans, so Mexicans got offended. Muslims got offended, a lot of different people around the world got offended by Donald Trump. So, he is an example of someone that is not politically correct at all.

Political Correctness can also have a bad side. It can make people feel afraid to say what they actually think. Like my grandma for example. She doesn't know the politically correct term for a certain people. She might innocently say the word Indian, she might call the First Nations people, Indians. So sometimes politically correctness can be bad because a person's intentions are good, but they just don't know. They aren't educated."[6]

And that's the problem! Who "educates" us on what's "politically correct" or not? What gives "them" that right "over" the rest of us? And whatever happened to "free speech" by the way? This is crazy, let alone unconstitutional! I mean, you can call it "political correctness" all you want, but in reality, all it is, is a slick "relabeling" by the "Elites" to "ban free speech" so they can control it all! That's all it is in a nutshell folks! You can call it "political correctness" all you want, you are just simply "banning" "free speech." You are controlling "information," "ideas," and "thoughts." These "Elites" are simply hiding behind this term to simply "control" the "narrative" in public, in print, whatever, right down to what "you can" or "cannot" "say" "write" "share" according to "them" and "their" agenda. Then they pitch it in a way that makes us think it's all for our good and the good of others, so nobody gets "offended." Yeah right, but at what cost? The cost again, is simply "free speech." And now it is being "banned" under the guise of "political correctness!" Very slick and deceptive! And this deceit has been going on for quite some time now. In fact, for so long, that it's now moved on to the second stage of attack on "free speech" and "control" of all "information," "ideas," and "thoughts."

That is with the new term out there called "Cancel Culture." First, it was "political correctness," now it's "cancel culture." Stage one, stage two. And in this second stage called "Cancel Culture" they're not only "banning" or "restricting" "free speech" so as not to supposedly "offend" others. But now they are "enforcing" this "banning" of "free speech" of those who persist in sharing their own free independent ideas or individual thoughts that the "Elite" doesn't want others to receive or goes against their preprogrammed narrative by literally "canceling" them out of existence altogether from the public arena. Don't believe me? As wild as it sounds, here's just a few examples of this "new trend" where people's voices are being "canceled" out.

What is Cancel Culture?

Dr. Tina Sikka, Lecturer of Media & Culture Newcastle University:
"Cancel Culture, I define as an act of public shaming that is based on something that either perceived or substantiated a social transgression of some kind, that hasn't been adequately addressed through traditional channels. So, things like racism, misogyny, ableism, transphobia, there is no way to get any kind of justice for comments that you might find offensive. Celebrities that have done something that people find kind of cringe-worthy and they would sort of as a hyperbolic statement say that they would be canceled. Then it kind of emanated into the larger conversation, it became involved in identity politics and so-called culture wars. And that usually means that anything that they have produced creatively, it's just not going to be viewed anymore. And there is a little bit of a media blackout."

Inside Edition: *"It's been called one of the greatest films of all times but in today's times, 'Gone with the Wind' has been pulled from the streaming service HBO Max. Long criticized for its racial stereotypes."*
"Gone with the Wind' is gone from HBO. It's one of the most popular films of all times. It's also been condemned for ignoring the horrors of slavery. The streaming service, HBO Max is now pulling the film."

Fox News: *"Where do you draw the line here? I'm told that no longer can you find on HBO 'Gone with the Wind' because somehow that is now somehow offensive. Where do you draw the line? Should George Washington and Thomas Jefferson and James Madison be erased from history? What about FDR and the internment camps?"*

Clip from movie 'The Help': *"It behooves me to inform you that you are fired Abilene."* The maid is told this as she is walking into the room with an armful of groceries.

Inside Edition: *"Twitter users today are now calling for Netflix to remove 'The Help' from its lineup, criticizing its 'white savior narrative.'"*

"Movies not to watch when trying to educate yourselves on racism: 'The Help,' goes one tweet."

"Other shows are coming under fire in the wake of the George Floyd protest movement. After 31 years 'Cops' has been canceled over concerns about racial depiction."

Looper Reports: *"Gina Carano won't return for the future seasons of the Pedro Pascal led series 'The Mandalorian' or for any other Star Wars project for that matter. On Wednesday, February 10, 2021, Carano, a former mixed martial artist who has become a recognizable face in the action and sci-fi genres over the years, was fired from 'The Mandalorian.' As reported by Deadline, a spokesperson for Lucasfilm confirmed on Wednesday evening that Carano wasn't just dropped from the first-ever live-action Star Wars series, she's also no longer to be considered for or offered employment with Lucasfilm in the future. The decision came after*

Carano continued to share on social media statements that Lucasfilm deemed, quote, 'abhorrent.' As Deadline details Carano's most recent controversial post on Instagram suggested that being a Republican in the US in 2021 is similar to being Jewish in Nazi Germany. This sparked outrage online and hashtag fire Gina Carano started trending on Twitter as folks tagged both Disney and Lucasfilm in thousands of tweets in an effort to remove her from 'The Mandalorian.'"

Fox News: *"Cancel Culture is claiming a new victim. Teen Vogue editor Alexie McCammond was forced out just days before starting her new job, after staffers revolted over derogatory tweets she wrote ten years ago when she was a teenager."*

Sky News Reports: *"The latest news is The Dr Seuss Books, a whole range of them are not going to be printed anymore because they offend people. Have you been able to work out what the problem is with, 'If I Ran the Zoo' or 'On Beyond Zebra' or 'Scrambled Eggs?'"*

Bella D'Abrera, Director of Western Civilisation Program, IPA: *"The problem is Cancel Culture on steroids. Cancel Culture today is absurd. It's particularly absurd in the US where it is just getting to new levels of extremes."*[7]

And unfortunately, it is spreading around the world! But I agree, this "cancel culture" mentality and behavior is insidious and absurd and dare I say dangerous and flat-out an assault on "free speech." They make it sound so good and imply it's good for the public so we can all get along, but all it is, is the strong arm of the "political correctness" crowd now applied to those who didn't "voluntarily" stop "saying" and "writing" or "reporting" things that you the "Elites" didn't want "said" or "wrote" or "reported" about, i.e., that were "politically incorrect." Cancel Culture has now become their strongarm weapon to literally "cancel" "silence" and "permanently remove" any voice that goes against their "controlled narrative" so that the public "only gets" what they "want us to get." Do you "see" the deception? Keep your glasses on! Because not so surprisingly, this "weapon" called "cancel culture" is also being used to encourage "scoffing" at the things of God, the Bible and Christianity, just like God warned about when we're living in the Last Days. Not only are Christian Books and the Christian Bible being banned by this "Cancel Culture" today, but of course, at the very same time just about any kind of book that "scoffs" or "promotes scoffing" at the Bible and "speaks negatively" about Christianity is glowingly "allowed." Let me give you just a couple quick examples of that.

"Will Amazon ban Christian books next? This continues a string of news stories about Amazon related to books by several authors that have been banned for various political reasons. Will Christian books be next?

Amazon chose to ban a book on 'conversion therapy,' and it will not be long before they begin to ban Christian books as well. For Christians, this should be alarming."

"A California bill bans the Bible. A California bill is so broad that it bans books, printed materials, and advertisements that provide information that a person facing unwanted same-sex attractions or gender confusion can change. California State Assembly member, Al Muratsuchi, said, 'The First Amendment does not prohibit banning fraudulent conduct. The faith community, like anyone else, needs to evolve with the times.'

"During his speech on the California bill at Google headquarters, LGBT activist Samuel Brinton reportedly said he wanted to 'figure out a way to stop' Pastors and Churches from offering counsel to help people change unwanted same-sex attractions. He went on to say, 'I may not be able to FIND every little camp...every pastor, but I can make it something that is culturally unacceptable.'

"The Bible specifically refers to homosexuality as **1 Corinthians 6:9-11** *states, 'And such were some of you,' thus clearly stating that through a personal relationship with Jesus Christ, people with same-sex attractions or who engaged in same-sex behavior can change. Advertising this passage or quoting it in a brochure would be considered fraudulent business practices under this bill, according to Liberty Counsel. This bill*

is an infringement on First Amendment rights and is a classic viewpoint discrimination. It declares certain kinds of speech as consumer fraud. Mat Staver, founder and chairman of Liberty Counsel, said, 'Their goal is to crush Christianity and any viewpoint that does not align with the state.'"

"Three Christian books banned from SRE curriculum in NSW. There has been quite some concern in Christian circles in my home State of NSW over the last few days, over bureaucratic action to 'ban' some books from being used in Special Religious Education classes.

"This sudden censorship was, frankly, astonishing. There had been no consultation or discussion on the matter.

"In a directive sent to NSW primary and secondary principals overnight, the Department of Education and Communities ('the DEC') said: 'It has come to the attention of the Department that resources being delivered to support Special Religious Education (SRE) in schools may be in conflict with departmental policy and legislative requirements.'

"No further explanation as to how the books were in conflict with the department's policy was given in the directive. The DEC instructed all principals to contact SRE providers to ascertain if the books are being used, and to direct providers to cease using them immediately.

"In a statement to ministers and SRE providers, Mr. Thorpe said 'We would like to remind churches and individuals that all SRE teachers must teach only the authorized curriculum resources.'

"Several SRE teachers and administrators were being approached by school principals to sign a document stating that they have not and will not use the three suspended books in their SRE classes.

"Michael Jensen says he is perplexed by the decision. 'I think this is an outrageous piece of book banning.'"

And I would agree. But my question is, "Again, how is this any different than the Fahrenheit 451 premise? How is this any different than what Hitler did with the German people when he literally "brainwashed" them into thinking that "Book banning" and "Book burning" was good for them?" In fact, let's remind ourselves of that dangerous historical event by Hitler.

Peter Black, senior historian, United States Holocaust Memorial Museum: *"In 1933, the Nationalist Socialist German Workers Party, called the Nazis for short, came to power in Germany and established a dictatorship under the leadership of Adolph Hitler. The Nazis intended to re-arm Germany and re-organize the German state, on the principal that the German ethnic group or race was superior to all others in Europe. They suppressed all dissent within Germany making it a crime to criticize the regime.*

The newly established Ministry of Propaganda and Enlightenment set up various chambers to control specific aspects of German culture, such as art, literature, theater, film, music, virtually all forms of entertainment and all forms of dissemination of news. In 1933, in April, Nazi German students decided to organize a nationwide book burning program to eliminate foreign influence, to purify German culture as they saw it."

William F. Meinecke, Jr. historian United States Holocaust Memorial Museum: *"So, you have committees of students meeting with professors together deciding what categories of books in these university libraries would count as un-German. They didn't see themselves as suppressing culture. They saw themselves as advancing Aryan German culture."*

Robert Behr, Holocaust survivor, born in Berlin, Germany: *"I remember very distinctly a conversation between my parents and some friends who were all shocked that a nation like the Germans, an educated, highly intelligent nation, would burn books. Books never hurt anybody."*

Peter Black: *"The event that the students planned occurred on May 10, 1933. In each German University City, thirty-four of them in all,* *thousands of people gathered together at a public place in which books that had been confiscated either by the students themselves or by the Nazi Party officials, often with the help of police, were brought and dumped in a pile. Student leaders exhorted their followers and the listening crowds to swear an oath by the fire, to destroy and combat subversive and un-German literature.*

William Meineke: *"For the national treason against our soldiers in World War I.*

Joseph Goebbels, the Propaganda Minister himself spoke at the book burning in Berlin: *"And the future German man will not just be a man of books, but also a man of character. And it is to this end we want to educate you."*

Azar Nafisi, author, 'Reading Lolita in Tehran': *"It is amazing to me the variety of books that were burned on that night and thereafter."*

William F. Meinecke: *"Among the authors whose books were burned were Ernest Hemingway, both Mann brothers, Thomas and Heinrich. Erich Maria Remarque who wrote the famous book 'All Quiet on the Western Front,' Helen Keller, Jack London."*

Ruth Franklin, literary critic and contributing editor, The New Republic: *"There is very little that unite them, all these books, except that they were considered dangerous by the Nazis."*

Peter Black: *"The grand total of the number of volumes, perhaps best estimates would be between eighty or ninety thousand volumes. For weeks afterwards books were confiscated from libraries, from bookshops, and from private collections.*

Azar Nafisi: *"The first thing every totalitarian regime does, along with confiscation and mutilation of reality, is confiscation of history and confiscation of culture. I think they all happen almost simultaneously."*

Ruth Franklin: *"Really all cultures are dangerous to a regime that fears the free flow of ideas. Because the literature in its most fundamental way is meant to forge connections among human beings."*

Azar Nafisi: *"Because you don't know where it takes you. Knowledge is always unpredictable, there is always a risk. It is like Alice jumping down that hole, running after that white rabbit, not knowing where she goes. And for tyrants, control is the main thing. They don't like this unpredictability. They don't want the citizens to connect to the unknown parts of themselves, of their past and to connect to the world."*

Ruth Franklin: *"For a totalitarian regime this is perhaps the most dangerous thing. Because these regimes are predicated on the idea that the people within them will resign themselves to thinking that this is all there is. And that there aren't any options."*

Robert Behr: *"How serious those warning signs were taken is exemplified by my mother, who when I asked her if we had to worry about a guy like Hitler, she said, 'No, we are living in a democracy. We have the protection of the police. Nobody's going to hurt us.' So, talk about warning signs, there were plenty of them. Did we take them seriously? My family didn't."*

Peter Black: *"In 1939 the Nazi regime initiated what became the Second World War. During the course of this war, the Nazis began to implement their population policy, a priority element of which was the annihilation of six million Jews on the European Continent, in a mass murder or genocide that we now call the Holocaust."*

Robert Behr: *"We never believed the Germans would stoop so low that they would implement the threats which one fanatic uttered. And so, our own life went from bad to worse and it culminated in July of 1942, when we were arrested and sent to a concentration camp. To make this clear, it was a life without hope."*

"Where books are burned, in the end people will be burned." Heinrich Heine, 1821 German poet.[8]

And that is exactly what happened. Folks, we better wake up and get our head out of the sand! History is repeating itself. Except this time, it is on a much grander scale. All they have done is change the terminology! Book burning, Book banning, political correctness, cancel culture, whatever you want to call it, it is all the same thing. Only this time, it's not just happening in Germany by one totalitarian dictator called Hitler. Now it is spreading across the globe and it's being done by a handful of "Aliens" I mean rich "Elites." But the premise is still the same. It's the beginning of a global dictatorship in which anyone who resists them or gets in their way of what they deem to be or desire as truth will eventually be taken out. First you burn the books, then you burn the people. When will we ever learn? But hey, good thing the youth of today and younger generations aren't falling for this book burning cancel culture mentality like Hitler did with the youth of his day. Oh, how I wish that were true. Unfortunately, these same "Elites" and all their Media Brainwashing Techniques have worked all too well on a whole new up and coming young generation that has actually been convinced that the "destruction" and "deletion" of any and all "dissenting" ideas or "independent" thoughts other than their own, is actually a good thing. Including going to the point of literally "burning" books, they disagree with and think others should not read, just like the youth did back in Hitler's days, only it's today! Watch this.

11alive.com: *"A meeting is going on right now at the Georgia Southern over book burning. It comes after students burned a Latina Author's book after she visited the school last week. Those students are going to be disciplined for this."*[9]

What you just saw, in my opinion, was a Modern-day Hitler Youth Movement taking place. Now, we don't call them Hitler Youth today, of course not! Why, haven't you heard? That is a "politically incorrect" statement. You cannot say that today, it might "offend" somebody! And whatever you do, don't say it's "Book burning." No! No! No! They are part of a whole new "Cancel Culture" movement that's only trying to "help people" not "offend" or "hurt others." Yeah right! What a modern-day deception! All they're doing is twisting their words in order to "silence" other people's words! It's Fahrenheit 451 becoming our reality. Hitler would have loved this movement! And the reason why is because it has been able to do what he wasn't able to do, that is, go beyond just one country and spread this "book burning" "information" and "thought control" movement across the planet! So, the next obvious question is, "How in the world did this happen? How did our nation turn so quickly from valuing the "rights" of "free speech" to the actual "restricting" and "canceling" of "free speech?" How did this mindset of a whole new generation arise on the scene where people not only "accept" but actually "promote" the idea that we "need" to let these "Elites" control all Books, Print Media, and overall information or thought for our own supposed good?"

Well, they simply did it by "controlling" the whole Education System, just like you saw in that new Fahrenheit 451 trailer! It's almost like somebody's following a script! Believe it or not, they admit this is precisely where the "brainwashing" needs to start! The Education System! You need to get them while they're young, like Hitler did! And if these same "Elites" who are "controlling" and "filtering" all the books and Print Media, and all the other forms of Media like Newspapers, Radio and Music as we saw, could "also" somehow "grab control" of the whole Educational System, then that means they could even "brainwash" a whole new generation into thinking that what they're doing is actually a "good" thing instead of a "bad" thing, and even get us to go along with it!

And that's precisely what they've done! Don't believe me? These "manipulated" byproducts of the Educational System today, who go along with the "Elites" agenda to "restrict" and "eliminate" "free speech," let alone any dissenting "idea," are not just real, but they have even been given a new name! Again, we don't call them Hitler Youth, that's not "politically correct." Have you already forgotten? Rather, we call them Snowflakes. You know, those young people our so-called Modern Education System is pumping out by the boatloads. Young impressionable minds that can no longer "tolerate" or handle any disagreeing opinion other than their own which has been shaped by the "Elites" and then "demands" the Government to "forbid" "ban" or "cancel out" any other views. You can actually see them in action here. Here's what your taxpayer dollars are producing in our Modern Education System!

Narrator: *"Thanks to major advances in technology, we now have access to tools that allow us to zoom in and list to the sound of an actual snowflake falling towards the earth. Preparing itself to experience a reality that is different from its own. Observe carefully and witness this once in a lifetime opportunity for yourself."*

Sitting in a crowd, listening for the 2016 election results there is a snowflake. When it is announced, that Donald Trump won the election, the snowflake opens its mouth to let out a blood curdling scream at the disappointment. Not one time did it scream, but over and over. The disappointment was just too much for it to handle.

Tucker Carlson: *"College students seem to get stressed out by finals, but if they are attending the University of Utah, you have an option. It's called the 'Cry Closet.' The 'Cry Closet' is the creation of an art student named Nemo Miller, and it exists for stressed out students. You can step into it for up to ten minutes for crying to help cope with the tests. The closet interior consists of fuzzy stuffed animals for maximum coping."*[10]

What has happened to our country? You don't like somebody else's "decisions" or "ideas" or "opinions", so you fall on the ground and scream your lungs out in utter despair and cry like a big baby? And yet, you are supposed to be an adult! And not only that, but you also literally can't handle any kind of stress, to the point where you need a "cry closet" full of stuffed animals to help you cope with life? What has happened to us? What a bunch of wimps! But folks, as crazy as this all sounds, and as unbelievable as this behavior is, it's not by chance. It's deliberately being done to them! This whole new Politically Correct, Cancel Culture, Book Burning generation called Snowflakes, are simply the byproduct of the whole new Modern Education system that's been hijacked by these same

"Aliens" I mean rich "Elites" who have not only taken over the various forms of media around the world as we saw with the Newspapers, Radio, Music and Books, but they have also simultaneously taken over the whole Education System to "brainwash" a whole new generation into accepting their total global control of all information we receive in order to "manipulate" us according to their nefarious global agenda! And as wild and conspiratorial as that sounds, let's now begin to take a look at how that actual "takeover" happened. And as we do, let's first go back and take a look how our Educational System "used to be" before it got hijacked by these "Aliens" I mean rich "Elites."

- The first colleges formed in America (123 out of 126) were formed on Christian principles.

- Up until 1900 it was very rare to find a university president who was not an ordained clergyman.

- The New England Primer, America's first textbook and used for 210 years taught the alphabet like this: A – In Adam's fall we sinned all. C – Christ crucified for sinners died. Z – Zacchaeus he did climb the tree our Lord to see.

- The 107 questions at the end of the New England Primer included questions like, "What offices does Christ execute as our Redeemer?" "How does Christ execute the office of a priest?" "What is required in the fifth commandment?" "What are the benefits which in this life do accompany or flow from justification, adoption, and sanctification?"

- George Washington made it crystal clear that American schools would teach Indian youths the "religion" of Jesus Christ and Congress assisted in doing so.

- In 1782, Congress had 10,000 bibles printed for use in schools.

- Dr. Benjamin Rush said, "The only means of establishing and perpetuating our republican forms of government is the universal education of our youth in the principles of Christianity by means of the Bible."

- Thomas Jefferson wrote the first plan of education for the city of Washington D.C. and adopted two textbooks, the Bible and Watts Hymnal, and hired clergymen to be the teachers.

- The 1854 edition of Webster's Dictionary had Biblical definitions, Bible verses, and Webster's own testimony of personally receiving Christ."

- America's first school was Harvard, founded in 1636 by Reverend John Harvard whose official motto was "For Christ and the Church." Harvard had several requirements which students had to observe, one of which was, "Let every scholar be plainly instructed and earnestly pressed to consider well, the main end of his life and studies is to know God and Jesus Christ, which is eternal life."

Boy, that's not what we have in Schools today, is it? Not even close! In fact, wait until you hear where the largest Church in America used to be.

David Barton, tour guide at the Capitol: *"This is a copy of what the first Bible printed in English in America looked like. This Bible was printed by the US Congress in 1782. In the record it says that this Bible was 'an neat edition of the Holy Scriptures for the use of our schools.' So, the first Bible, printed in English, in America, was printed by Congress, for the use of our schools, it's worse than that. On the front of the cover, it says that Congress resolved that the United States and Congress recommend this edition of the Bible for its inhabitants in the United States. So, the first Bible, printed in English, in America, was done by the guys that signed the documents, endorsed by Congress, and done for use in schools and we are going to be told that they don't want any kind of religion in education, they don't want voluntary prayer, no. It doesn't make sense. This document by itself is very significant.*

In 1830 Congress commissioned these four paintings over here to recapture what the official record says was the Christian history of the United States. In these four paintings you have a span of several hundred years. By taking you through chronologically, the first is Columbus landing in the Western World in 1492. They got out, they knelt down and had a prayer service. Do you see the cross they had? They named the land where they had landed, San Salvador. Meaning Holy Savior, which tells you something of the thinking that was going on then.

You come back over my shoulder here, this is the baptism of Pocahontas in Jamestown in 1613. Over here, the fourth painting is in 1620. This is the embarkation of the pilgrims coming to America. You see them gathered around the Bible at the prayer meeting they are having. If you just take those four paintings there, those four paintings, in this great secular hall of government, represent 2 prayer meetings, a Bible study, and a baptism, which is not bad for a secular building. As a matter of fact, you are standing in what, in 1857, was the largest church in the United States. The US Capitol.

Back on December 4th in 1800 members of Congress, members of the Senate, Thomas Jefferson was over the Senate, John Trumble on the House, they decided to turn this Capitol building into a church building. Starting on Sunday they started having services in the Capitol. Six weeks after that Thomas Jefferson became president of the United States. During his eight years of being president, he came to church here at the US Capitol, listened to the services here at the Capitol and being Commander in Chief, he decided he could help the worship here at the Capitol. He ordered the Marine Corp band to come play at the worship service. Now that would be rather cool to have the Marine Corp band play the music at the worship service.

That church went on for the better part of the century and by 1857 there were two thousand people who went to church in the Hall of the House of Representatives. In addition to that there were four other churches that met at the Capitol. First Congregational, this was their church home, as was First Presbyterian, as was Capitol Hill Presbyterian. Churches met here, there was nothing secular or seemed to be secular about this building until the last 30, 40, 50 years."[11]

Until it got hijacked by the "Elites" as we will soon see. But wow, they don't tell you that interesting piece of American History in the Media or School System today, do they? Not at all! I wonder why? Well, we'll find out here in a second. But if you want even more proof that our Founding Fathers wanted the Bible and God's principles to be integrated and taught in our Education System, let me give you just a few more quotes from them.

John Adams: *"The general principles on which the fathers achieved independence were the general principles of Christianity. Without religion, this world would be something not fit to be mentioned in polite company: I mean hell. The Christian religion is, above all the religions that ever prevailed or existed in ancient or modern times, the religion of wisdom, virtue, equity and humanity."*

John Quincy Adams: *"My hopes of a future life are all founded upon the Gospel of Christ. In the chain of human events, the birthday of the nation is indissolubly linked with the birthday of the Savior. The Declaration of Independence laid the cornerstone of human government upon the first precepts of Christianity."*

Patrick Henry: *"Being a Christian is a character which I prize far above all this world has or can boast. Righteousness alone can exalt America as a nation. The great pillars of all government and of social life are virtue, morality, and religion. This is the armor, my friend, and this alone, that renders us invincible."*

John Jay: (Original Chief Justice of the U.S. Supreme Court) *"Mercy and grace and favor did come by Jesus Christ. By conveying the Bible to people, we certainly do them a most interesting act of kindness. The most effectual means of securing the continuance of our civil and religious liberties is always to remember with reverence and gratitude the source from which they flow. The Bible is the best of all books, for it is the word* *of God and teaches us the way to be happy in this world and in the next. Continue therefore to read it and to regulate your life by its precepts. Providence has given to our people the choice of their rulers, and it is the duty as well as the privilege and interest of our Christian nation, to select and prefer Christians for their rulers."*

Benjamin Rush: (Father of Public Schools Under the Constitution) *"The Gospel of Jesus Christ prescribes the wisest rules for just conduct in every situation of life. Happy they who are enabled to obey them in all situations! The great enemy of the salvation of man (Satan) in my opinion never invented a more effective means of limiting Christianity from the world than by persuading mankind that it was improper to read the Bible at schools. Christianity is the only true and perfect religion; and in proportion as mankind adopt its principles and obey its precepts, they will be wise and happy. The Bible should be read in our schools in preference to all other books."*

Noah Webster: (Schoolmaster to America) *"The religion which has introduced civil liberty is the religion of Christ and His apostles. This is genuine Christianity and to this we owe our free constitutions of government. The moral principles and precepts found in the Scriptures ought to form the basis of all our civil constitutions and laws. I am persuaded that no civil government of a republican form can exist and be durable in which the principles of Christianity have not a controlling influence."*

George Washington: *"You do well to wish to learn our arts and ways of life, and above all, the religion of Jesus Christ. These will make you a greater and happier people than you are. While we are zealously performing the duties of good citizens and soldiers, we certainly ought not to be inattentive to the higher duties of religion. To the distinguished character of Patriot, it should be our highest glory to add the more distinguished character of Christian."*

Congress U.S. House Judiciary Committee 1854: *"Had the people, during the Revolution, had a suspicion of any attempt to war against*

Christianity, that Revolution would have been strangled in its cradle. In this age, there can be no substitute for Christianity. That was the religion of the founders of the republic, and they expected it to remain the religion of their descendants."

Sounds to me like the Founding Fathers of America clearly wanted, planned, and even provided for the Bible and God's principles to be taught in our Education System and to be promoted throughout our country. In fact, we even had Presidents who did these kinds of godly Biblical things while they were still in office. Check this out!

David Barton, tour guide at the Capitol: *"Do you see that statue to the left of the door over there, that white marble statue? That is President James A. Garfield. President Garfield was one of the young Major Generals in the Civil War. He was a war hero. He became Speaker of the House, he became the 20th President of the United States and by the way, that man founded Howard University, General Howard took it over after he founded it. Just a really cool guy. But what we never hear about that President of the United States is that he was a minister during the second great awakening.*

This is actually one of his letters signed, James A Garfield, 1858, and in this letter President Garfield recounts that he had just finished preaching a revival service where he had preached the gospel nineteen times in the revival. He said as a result of his preaching, 34 folks came to Christ and he baptized 31 of them. Now that doesn't sound like a typical Presidential activity today but that is what we used to do with Presidents in the past. Again, you can walk through and see that statue and say there's a President, but you would never think that there is a minister. We have so compartmentalized Christianity into such a small box that we don't realize that our military leaders, our ministers, our educators, our Presidents, used to be ministers. I would say about one-fourth of these statues are ministers of the gospel."[12]

Wow! Why don't they share that in our Education System today? Well again, we'll get to that in a second, but let's first answer the obvious question, "Why were the Founding Fathers so adamant on making sure we had the Biblical teachings and Christianity and God's Holy principles taught in our Education System?" Because one, they wanted the people to know clearly where our rights and freedoms as Americans came from. They come from God, not the Government. This is what our country was built upon. I didn't say that they did, it's in the Declaration of Independence.

"When in the Course of human events, it becomes necessary for one people to dissolve the political bands which have connected them with another, and to assume among the powers of the earth, the separate and equal station to which the Laws of Nature and of Nature's God entitle them. A decent respect to the opinions of mankind requires that they should declare the causes which impel them to the separation.

We hold these truths to be self-evident, that all men are created equal. That they are endowed by their Creator with certain unalienable Rights, that among these are Life, Liberty and the pursuit of Happiness."

Secondly, our Founding Fathers made sure that the Bible and Christianity and God's principles were woven into our Education System not only so people would know where our rights and freedoms come from, that is God, our Creator, not the Government, but also "to keep our hearts in check" "morally" so we would not get tricked into going back to a tyrannical form of Government. You see, the Founding Fathers knew that the Bill of Rights and Constitution were only an outward shell that could only provide general direction for the Government. But they could not tame the internal heart of man, to keep him under control "morally." The taming of the heart, they believed, was up to the Bible! That's why they wanted to ensure that the Bible was taught throughout our Education System! They knew that if America's heart got inwardly rotten again "morally" then the Bill of Rights and Constitution could not stop us as a people from going backwards again, back to tyranny and oppression that the Founding Fathers escaped from back in England. So, they wanted to make sure that all people in America "could" and "would" study the Bible to stay morally in check to prevent us from going back to tyranny!

But look at us today! We are no longer taught these facts about the true intentions of our Founding Fathers of having the Bible, Christian teachings, and Godly principles permeating throughout the Education System for stated reasons, but now we have an Education System that says the exact polar opposite! That there is no Creator, and that our rights come from the Government, not from God, and that even our Constitution, which was built on Biblical principles of freedom, is actually now a dangerous document. In other words, it's not "politically correct" today! Don't believe me? Recently a Publishing company actually put a "warning label" on the Constitution! And I quote:

"Wilder Publications warns readers of its reprints of the Constitution, the Declaration of Independence, Common Sense, the Articles of Confederation, and the Federalist Papers, among others, that this book is a product of its time and does not reflect the same values as it would if it were written today."

The disclaimer goes on to tell parents that they *"might wish to discuss with their children how views on race, gender, sexuality, ethnicity, and interpersonal relations have changed since this book was written."*

Excuse me? A warning label on the Constitution of the United States of America? Folks, that's not only ludicrous, but it's also blasphemous to the original intent of our Founding Fathers. And then on top of that, we even have the long-standing devious lie that's also spread throughout the minds of the American people and that is the lie that our Founding Fathers did "not" want the Bible or Christian teachings, or God's principles taught in our schools! No! They say, rather, our Founding Fathers wanted, "Separation of Church and state," which basically has come to mean today, "Keep God, Christianity, and the Bible out of our Education System, and our lives, at all costs." Now again, this is not only ludicrous, as we saw based on the actual historical statements made by our Founding Fathers, but not so surprisingly, this "lie" is simply the twisting of a historical phrase by another one of our Founding Fathers, that of Thomas Jefferson. The next time someone spouts off the phrase, "Separation of Church and State" and they say that phrase proves that our

Founding Fathers wanted to get rid of Christianity and the Bible and God's principles, you need to challenge them. Ask them this. "Where does that phrase, "Separation of Church and State" appear in the Bill of Rights? Where does it appear in the Constitution?" Answer? It's not there in either one of them! Rather, it came from a letter from Thomas Jefferson in 1802 to the Danbury Baptists of Connecticut who were concerned that another popular Christian denomination at that time called the Congregationalists were going to become the official Christian denomination of the United States.

Now, notice the whole time the context is Christians and Christianity. Not the exclusion of it let alone the acceptance and promotion of Buddhism, Islam, Eastern Mysticism and everything else under the sun! No! When our Founding Fathers talked about "religion," they were talking about Christianity! And so, in that letter, Jefferson simply referred the Danbury Baptists and their "concern" back to the first amendment that reads, "Congress shall make no law respecting an establishment of religion or prohibiting the free exercise thereof." In other words, he had full confidence in the "original intent" of the first amendment that it would ensure that no "Christian denomination would be chosen to be the official Christian denomination of the United States." That's it! That's what he meant by the phrase, "Separation of Church and State." And again, it did not have anything to do with keeping Christianity, God and the Bible and Christian teachings out of our Education System or society today! We know that not only by the Historical context of that letter from Jefferson, but again, also from the very words of our Founding Fathers that clearly stated they wanted to "keep" the Bible, Christian teachings, and God's principles in our Schools and society for all time!

And so, the next common-sense question is, "How in the world did this happen? How did our Nation that was built upon Biblical principles of "freedom," and our "rights" that come from God, get so twisted around that we actually now "forbid" the Bible in our whole Education System? Well again, it's been hijacked by a group of "Aliens" I mean rich "Elites" who are using the Education System to promote "Humanism" in order to take us back to a tyrannical form of Government to "control" us. And for those of you who may not know what Humanism is, it's the worldview where "man" is basically the center of all things, not God. And so right out of the gates, you can see this belief system is diametrically opposed to the Biblical teachings that "God" is the center of all things and as our Founding Fathers believed, our rights come from Him, "The Creator" not "man." So, these "Elite's" first step was to twist this "protective" Biblical belief on its head in order to "brainwash" the American people into going back to tyranny by conditioning them to think our rights came from "man" or the Government, not God. And they did it with the Education System. But don't just take my word for it, let's listen to theirs. Let's take a look at their own Humanist beliefs from the Humanist Manifesto I and II, and let's see if their "man-centered" anti-Biblical belief system is being taught in our Educational System today.

- Faith in the prayer-hearing God, assumed to live and care for persons, to hear and understand their prayers, and to be able to do something about them, is an unproved and outmoded faith.

- We find insufficient evidence for the belief in the existence of a supernatural. We begin with humans not God.

- We do not accept as true the literal interpretation of the Old and New Testaments.

- We include a recognition of an individual's right to die with dignity, euthanasia, suicide, birth control and abortion.

- We believe that intolerant attitudes, often cultivated by orthodox religions and puritanical cultures, unduly repress sexual conduct. Divorce should be recognized. The many varieties of sexual exploration should not themselves be considered evil.

- We oppose any tyranny over the mind of man to shackle free thought. In the past such tyrannies have been directed by churches and states attempting to enforce the edicts of religious bigots.

- Promises of immortal salvation or fear of eternal damnation are both illusory and harmful.

- Salvationism, based on mere affirmation, still appears as harmful, diverting people with false hopes of heaven hereafter. No deity will save us; we must save ourselves.

As you can see, all those Humanist beliefs, every single one of them, are now being taught throughout our Education System today, instead of the Biblical teachings that our Founding Fathers wanted! And lest you doubt they've hijacked the schools to promote this lie, it was none other than John Dewey, the man who is considered to be the "father" of our so-called Modern Educational system, who not only was an original signer of the Humanist Manifesto and celebrated today and given high honors by the NEA, or the National Education Association, but he is the one who wrote the theory of education we use today in our schools based on evolution and Humanism! And from there, these same Humanists went on the attack and forced the Bible, prayer, and even the Ten Commandments to be removed from our school system, court rooms, and Governmental Institutions under the guise that "religion" and "education" do not mix. You know, the lie of "Separation of Church and State!"

And yet, Humanism itself was declared as a "religion" by the Supreme Court back in the early 1960s! And can you guess whose "religion" is being "taught" in schools today? That's right, Humanism! Anybody seeing the irony here? So why would these Humanists want to take over the Educational System, Court Rooms, and Government, with their "anti-God" teachings? Because again, they're taking over the hearts and minds of the America people for "control!" They're "mesmerizing the minds of the masses" to take us back to tyranny! Again, our country was built upon the Biblical teachings that our "rights" and "freedoms" come from God, not the Government. And as long as we continue to be "educated" into these Biblical "truths" that we Americans "hold self-evident" as stated by our Founding Fathers, then we will always be protected from going back to a tyrannical form of Government. Therefore, if someone is ever going to "trick" us into going back to tyranny, then the first thing they have to do is remove the Biblical "protection." That's why these "Elites" are today "reeducating" and "mesmerizing" the minds of the American people "away" from the Bible, Christian teachings, and God's principles" in the Education System and supplanting it with Atheism, Humanism, and an evolutionary world view that says our rights come from "man," not God. And that's not just what they've done as we clearly saw tracing the trail of historical evidence, but believe it or not, they even admit that this was their plan from the very beginning! They wanted to hijack the American School System in order to "mesmerize the minds of the masses" or the American people, into Humanism man-centered teaching, and away from God! Don't believe me? Here are their own words admitting it!

- *"(Our) great object was to get rid of Christianity and to convert our Churches into halls of science. The plan was not to make open attacks on religion...but to establish a system of state schools, from which all religion was to be excluded...and to which all parents were to be compelled by law to send their children. For this purpose, a secret society was formed, and the whole country was to be organized."* **Orestes Brownson (1803-1876)**

- *"What the Church has been for medieval man, the public school must become for democratic and rational man. God would be replaced by the concept of the public good."* **Horace Mann (1796-1858)**

- *"There is no God and there is no soul. Hence, there is no need for the props of traditional religion. With dogma and creed excluded, then immutable truth is also dead and buried. There is no room for fixed, natural law or moral absolutes."* **John Dewey (1859-1952), the "Father of Progressive Education;" co-author of the first Humanist Manifesto and honorary NEA President**

- *"Education is thus a most powerful ally of humanism, and every American school is a school of humanism. What can a theistic Sunday School's meeting for an hour once a week and teaching only a fraction of the children do to stem the tide of the five-day program of humanistic teaching?"* **Charles F. Potter, Humanism: A New Religion (1930)**

- *"I think that the most important factor moving us toward a secular society has been the educational factor. Our schools may not teach Johnny to read properly, but the fact that Johnny is in school until he is sixteen tends to lead toward the elimination of religious superstition."* **Paul Blanshard, "Three Cheers for Our Secular State," The Humanist, March/April 1976**

- *"We must ask how we can kill the God of Christianity. We need only to ensure that our schools teach only secular knowledge. If we could achieve this, God would indeed be shortly due for a funeral service."* **G. Richard Bozarth, "On Keeping God Alive," American Atheist, November 1977**

- *"I am convinced that the battle for humankind's future must be waged and won in the public-school classroom by teachers who correctly perceive their role as proselytizers of a new faith: a religion of humanity. These teachers must embody the same selfless dedication as the most rabid fundamentalist preachers, for they will be ministers of another sort, utilizing a classroom instead of a pulpit to convey humanist values in whatever subject they teach, regardless of educational level – preschool, day care or a large state university. The classroom must and will become an arena of conflict between the old and the new – the rotting corpse of Christianity...and the new faith of humanism."* **John J. Dunphy, "A New Religion for a New Age," The Humanist, January/February 1983**

- *"Every child in America entering school at the age of five is mentally ill, because he comes to school with certain allegiances toward our founding fathers, toward our elected officials, toward his parents, toward a belief in a Supernatural Being, toward the sovereignty of this nation as a separate entity. It is up to you teachers to make all these sick children well by creating the international children of the future."* **Harvard Professor of Education and Psychiatry, 1984**

How can you get any clearer than that? Schools and the Educational System are the means of which these "Elites" will use to "brainwash" a whole new generation of Americans into "Humanism" and way from "God" in order to trick us into going back into a tyrannical form of Government, where "man" or these "Aliens" I mean rich "Elites" will dictate our very "rights" and "freedoms" and "free speech" and "beliefs" and even "behavior," not God. And unfortunately, their plan has worked all too well because as we saw, these new "international children of the future" have been "created" via the Education System, known as "Snowflakes." They really believe that "man" or the Government should dictate every facet of our lives and even "punish" or "ban" or "cancel out" anyone who disagrees with that! Isn't that wild?

That's not "freedom" it's "slavery." Our Founding Fathers must be rolling over in their graves! But as you can see by their own words, Humanists admit they hijacked our Education System with their man-centered belief system and are now using the schools to turn kids away from God and the Biblical principles of freedom in order to "brainwash" and "indoctrinate" them in Humanism so these "Elites" can rule over us. In fact, if you want even more proof that the schools are doing this to the minds of a whole new young generation of impressionable children, here's one lady who used to work for them, but now she's blowing the whistle on them! She admits, they not only "don't" want to teach your kid about education, you know, reading, writing, and arithmetic, but to indoctrinate them into a Humanist mindset against God and Biblical principles of freedom!

Charlotte Iserbyt: *"My name is Charlotte Iserbyt, and I served as senior policy adviser in the United States Department of Education under the Reagan administration. During which I had access to all the most important documents for the restructuring for not only the American education but for the global education. I am also the author of 'The Deliberate Dumbing Down of America' which gets into all of this, into the background of what I saw not only in the Department of Education but as a local school board member.*

Just to give you an idea of how blatant they are, Benjamin Bloom said the purpose of education, and often says parents really listen to us, you think the real purpose of education is reading, writing and arithmetic? The purpose of education is to change the thoughts, actions, and feelings of students. And then he goes on and he says he defines good teaching, and this is even worse from the parental standpoint, as challenging the students fixed beliefs.

And then in some of his works that I have, it's all in my book, say he can take a student, it's through challenging, you go up against them and you challenge them, then he goes on in one part of the taxonomy that is in my book, he says he can take a student from here to there, from a belief in God or his country, whatever, to being an Atheist and not believing in his country in one hour. Oh yes, he does, I've seen it, I've seen it with young people."[13]

Wow! Talk about shades of Hitler! So much for reading, writing, and arithmetic! That's not education, that's indoctrination! That's "brainwashing." They admit they're using the whole Education System to "brainwash" kids away from "God" and "Biblical principles" into Humanism and Atheism? And they can do it in one hour! Why? Because that's the first step into tricking a "free" people of a free nation into going back to "slavery" and "tyranny!" And as she noted, this wasn't just going on in America, but what? Internationally! In "all the schools" across the globe! And if you want further proof of that global invasion, then look no further than who controls all the "content" of all the Books that are being used to "educate" or should we say "indoctrinate" or "brainwash" kids around the whole planet! Just like we saw with Secular Book Publishers, the Publishers of "textbooks" used in schools today also call themselves "The Big Five."

- **#1 McGraw-Hill Education**
- Annual revenue: $1.7 billion
- Notable imprints: Glencoe/McGraw-Hill, Macmillan/McGraw-Hill, McGraw-Hill Higher Education, New editions of test prep books (SAT and ACT) and elementary school math textbooks

- McGraw-Hill Education should ring a bell for anyone who's experienced the magic of the American public school system. As one of the "big five" educational publishers, McGraw-Hill has long played a major part in providing textbooks and other curriculum materials for the K-12 set. And while much of its current strategy focuses on digital content and technological learning solutions, it still has a serious horse in the race of textbook publishing.

- **#2 Scholastic**
- Annual revenue: $1.7 billion
- Notable imprints: Arthur A. Levine, Klutz Press, Orchard Books
- Another major publisher of educational texts as well as kids' books, Scholastic was founded in 1920 and has remained relatively intact — that is, unlike many other publishers. No one else owns it. Scholastic is the largest publisher and distributor of children's books in the world, with perpetual rights to many of the most famous children's and YA series of all time, such as Harry Potter (recently acquired from Bloomsbury) and The Hunger Games. Its annual revenue has averaged around $2 billion over the past couple of years, and it consistently publishes some of the most popular titles in children's literature.

- **#3 Cengage Learning**
- Annual revenue: $1.5 billion
- Notable imprints: Chilton, Education To Go, Gale, National Geographic Learning
- Cengage is an educational publisher that's on the rise, especially after its most recent fiscal report. Cengage's profits last year came largely from online textbooks and other course materials, similar to McGraw-Hill. However, despite its present business model being digitally based, Cengage is still linked to traditional publishing in the form of a library division called Gale. Between the research and academic publishing that Gale conducts and the thorough online resources that Cengage supplies, we can expect to see Cengage rise in these rankings over time.

- **#4 Houghton Mifflin Harcourt**
- Annual revenue: $1.4 billion
- Notable imprints: Clarion, Graphia, John Joseph Adams Books, Sandpiper, Elementary school textbooks in all subjects
- Officially formed in 2007 after a merger between (You guessed it!) Houghton Mifflin and Harcourt, Houghton Mifflin Harcourt is the second of the "big five" educational publishers. Though it was formerly a subsidiary of Education Media and Publishing Group, HMH now owns itself and all its imprints, which include Holt McDougal and Riverside Publishing. Like McGraw-Hill and Cengage, HMH serves the K-12 education market and sees most of its profits from that.

- **#5 Pearson Education**
- Annual revenue: $1 billion
- Notable imprints: Adobe Press, Heinemann, Prentice Hall, Wharton Publishing
- Pearson PLC was once a part-owner of Penguin Random House, but its Pearson Education division is limited to academic texts. This is the third of the "big five" educational publishers and, similar to its fellows, Pearson Education has made great digital strides to stay at the top of the market over recent years. PE has recently partnered up with language learning service DuoLingo to reach more consumers, and it's also established a website-operated textbook rental service. Indeed, like other educational publishers, textbooks remain a cornerstone of its business.

And so, as you can see, a handful of "Aliens" I mean rich "Elites" own and control all the "textbooks" for the "education" of the whole planet, just like they "control" all the "regular books" on the planet! Anybody seeing a pattern here? And if you want even more proof that this is being done on a "global" scale, then consider this next thread of historical evidence about of our Education System. It's being dictated by the United Nations and the "Elites." The Education Plan designed by the "Elites" to "brainwash" the whole planet is simply called the "World Core Curriculum." George Bush Senior brought it into our Education System and simply renamed it "America 2000." Then, it was picked up by Bill Clinton and renamed "Goals 2000." Why did he call it Goals 2000? Because the goal was to get this "world curriculum" implemented into the American schools by the year 2000. It was all coming from the United Nations and the "Elites." Then it transferred hands from Bill Clinton to George Bush Junior who simply changed the name to "No Child Left Behind." And then the same deceptive relabeling technique continued on with Obama and he called this global brainwashing curriculum for the planet, "Common Core." But it is all the same thing.

Now you know why our kids and society have been getting dumbed down more and more and are clueless about our true history, including the beliefs of our Founding Fathers and where our "rights" and "freedoms" truly come from. God, not the Government. It's being done on purpose. These "Aliens" I mean rich "Elites" have "hijacked" and now totally "control" our Education System and are using it to "prepare" kids, literally "brainwash" them, and "mesmerize their minds" into "thinking" what the "Elites" want them to think in order to enslave them and the whole planet into accepting and going along with and even promoting their nefarious agenda. Pretty slick, isn't it? And speaking of deception taught in Schools today, you might be thinking, like a good little Snowflake, "Well there's nothing wrong with having Humanism, Atheism, Evolution and different views of origin taught in our Education System today. I mean, kids can make up their own minds based on the facts, right, what's right or not? Isn't that what schools are all about? Teaching kids to be "critical thinkers?" Well, it used to be, but not anymore. Now our so-called Modern School System exists not to teach kids to be "critical thinkers," but rather "what to think." That's not "education" that's "indoctrination." And they're doing that by "forcing" students to have only "one" view of origins in schools, that of Atheism, Humanism, Evolutionary ideals, and purposely "excluding" all mentioning of the Biblical view of origins that our Founding Fathers believed in and wanted for the American people to teach us that our "rights" and freedoms" come from God our Creator, not man. By the way, if you want more facts about how Humanism, Atheism and Evolution is a lie, then get our study entitled, "The Witness of Creation" so you can get equipped with the facts that demonstrate clearly just how bankrupt, untrue, and even unscientific these belief systems really are. But again, this is the hypocritical behavior of our Education System today. The irony is the whole initial original argument from these Humanists to get their man-centered atheist teachings including Evolution into our schools was that it was "unfair" to have just "one" view of origins taught in our Education System. This was their original argument during the Scopes Monkey Trial as seen here.

"Until the 1990s no trial in American history attracted more attention than this one did, that is, The Scopes Monkey Trial. It was held in 1925 in Dayton, Tennessee, and it accused a teacher of violating a state law that banned the teaching of human evolution.

Although the teacher had not taught biology and could not even remember for sure whether he had discussed evolution while substituting for the regular teacher, he agreed to be arrested and stand trial anyway.

For the contest, the ACLU brought in several big-city attorneys, including the famed criminal lawyer and atheist Clarence Darrow from Chicago. And to assist the prosecution, the World's Christian Fundamentals Association secured the services of William Jennings Bryan of Nebraska, a three times candidate for President and antievolutionist.

It was until this trial that most public schools taught the divine creation theory. Students in those days studied the facts of science and were told that the evidence indicates there is a Creator who designed the universe and plants and animals. But that was soon to change.

The trial lasted eight days and as expected, it ended in a conviction for the young teacher, whose own attorneys even conceded his guilt.
However, in that famous trial, atheist lawyer Clarence Darrow said, 'It is bigotry to only teach one view of origin.' He said, 'Students should be taught both the creation and evolution theories.'

So, in the 35 years following that trial, the theory of evolution was not only taught more and more, but it was being presented as if it were a proven scientific fact. Until today, we actually have this trial in reverse. Evolution is now the only theory that is taught."

Gee, that sounds hypocritical! But what most of us "also" don't realize about the Scopes Monkey Trial is that it seemed to be a watershed event that forever changed our nation for the worse. And that's because even though evolutionists lost that famous case, a momentum for evolutionary teachings began to take a foothold in our Education System. And so, over the next 35 years, the theory of evolution began to be taught more and more in our schools. Until finally, in the early 1960s, prayer and Bible reading were eventually taken out and evolution was put in permanently. And to help the American people to buy into these "lies" that were introduced in our Education System in order to "brainwash" us "away" from God and the knowledge of where our "rights" and "freedoms" come from. Hollywood, another form of media that we'll get into next, began to "mesmerize the minds of the masses" into thinking this "Scopes Monkey Trial" was a "good" "liberating" thing for the American public. They did this with a movie called, "Inherit the Wind." But the problem is, once again, these "Hollywood Humanists" lied to advance their freedom killing agenda. Let's take a look at that deception!

A clip from "Inherit the Wind."

In a courtroom scene, Spencer Tracy is speaking. *"Darwin took us forward to a hilltop from where we could look back to see the way of which we came. For this insight and for this knowledge we must abandon our faith in the pleasant poetry of Genesis."*

Narrator: *"One of the most enduring films of its era, 'Inherit the Wind' has been universally hailed as a brilliant recreation of a fascinating episode in American history. Popular reviewer Leonard Maltin called it an absorbing drama and based on the notorious Scopes Monkey Trial of 1925. And then went on to say the issue is real and still relevant. The issue, of course, is evolution. And undergirding it is the classic struggle* between religion and rationalism. Whether God's Word or the suppositions of science should provide the foundation upon which men and women order their lives.

Had 'Inherit the Wind' been faithful to the facts concerning this quote, 'fascinating episode in American history' end quote, Hollywood would be immune to the charges of bias on this important issue. The truth is however that the drama played out on the big screen bore little resemblance to the actual trial. Consider just a few of the film's many distortions and the bias they reveal.

The movie opens with a rather ominous rendition of the song, 'Old Time Religion'. As drums pound, sinister-looking men begin to gather together. The focus of their dark conspiracy is revealed as they enter the local high school and burst into a science class. There John Scopes is found passionately teaching evolution to his eager students."

Teacher John Scopes: *"We are finished with our lesson for today. We will continue our discussion of Darwin's theory of the descent of man."*

Narrator: *"He is promptly thrown into jail where he remains throughout the course of the trial. Help then arrives in the persons of two smart tough-talking agnostics, writer HL Mencken and trial lawyer Clarence Darrow (played by Spencer Tracy). Scopes if for the moment saved from the fundamentalist Christians who gather outside his jail to throw things and breathe threats of a lynching.*

In reality John Scope's classroom was never invaded. His arrest was a sham perpetrated by pro-evolution forces and he never spent a moment in jail. A math teacher and football coach, Scopes had spent two weeks substitute teaching in a biology class where he simply helps students review for a final exam. The trial came about as a result of the ACLU's desire to test a state law forbidding, not the teaching of evolution, per se, but only the theory that man had descended from lower animals. Scopes reluctantly volunteered to be their guinea pig and in return had his tuition covered for graduate school.

As for the mean-spiritedness and hypocrisy of a Bible-believing town folk a theme the film pounds home time and again, visiting reporters in fact noted the kindness and 'extreme courtesy of the town's citizens.' Clarence Darrow himself observed that he had 'been treated better, kindlier and more hospitably than I fancy would have been the case in the North.'

Perhaps even more dishonest than these distortions of historical records is the film's portrayal of the renowned lawyer and three-time nominee for President William Jennings Bryan, as the prosecutor in this case. It fell to Bryan to defend the Biblical account of man's origins. The film uses the epic courtroom battle between Bryan and Darrow as a dramatic device to illustrate the struggle between faith and suppositional science. Guess who's made to look like an idiot?"

Bryan: *"The Bible is enough to satisfy me. It is enough!"*

Darrow: *"It frightens me to think of the state of learning in the world if everybody had your driving curiosity."*

Narrator: *"Throughout the film, Bryan is portrayed as a stupid close-minded, hypocritical egotist. One minute he's shown cruelly intimidating Scope's fiancé on the witness stand. The next using legal pretense to block scientific testimony he fears will destroy his case."*

Bryan: *"I refuse to go along with these agnostic scientists to employ this courtroom as a sounding board. There's a platform from which they shout their heresies into the headline."*

Narrator: *"He's portrayed as blindly intolerant when he states his opposition to the use of Darwin's book in public schools and then admits he's never read it."*

Bryan: *"I am not the least interested in the hypothesis of that book! I never read it and I never will!"*

Narrator: *"Under cross examination he prudishly states that his belief that sex is inherently sinful."*

Darrow: *"What is a Biblical evaluation of sex?"*

Bryan: *"It is considered original sin."*

Narrator: *"And finally sensing defeat, he insists upon giving a final desperate and ultimately pathetic summation and then slumps to the floor in the throes of death. The fact is, the real William Jennings Bryan was almost universally recognized as a brilliant gracious Christian gentleman. One expert on the Scopes trial, despite the fact that he disagreed with Bryan's religious and scientific views, was nevertheless forced to admit 'as a speaker, Bryan radiated good-humored sincerity.'*

In personality he was forceful, energetic and opinionated but genial, kindly, generous, likeable, and charming. He showed a praise-worthy tolerance towards those who disagreed with him. Bryan was the greatest American orator of his time and perhaps any time. The fact is no woman ever took the stand during the trial. More importantly, court transcripts reveal that it was Darrow, not Bryan, who was at times rude, drawing even a contempt of court citation for repeatedly interrupting and insulting the judge."

The fact is it was also Darrow, not Bryan, who prevented the scientist from testifying. Bryan had been given the right to cross-examine and Darrow was so opposed to his experts being questioned that he never called them to the stand. The fact is, it was Bryan who introduced Darwin's book, then quoted from it as evidence in the trial. Transcripts again reveal that of the two lawyers it was Bryan who had the better command of the meaning and the mechanism of evolution.

The fact is, Bryan made no mention of sex during his testimony. Apparently, Hollywood couldn't resist introducing sex into the film as well as the implication that the Bible believing Christians are prudes. And finally, the fact is, Bryan gave no final speech. He did later submit a written summation that was characteristically clear and well-reasoned. He died in his sleep of an unknown cause five days after the conclusion of the trial.

Today some thirty years later, television docudramas and movies, like JFK, are criticized for at times playing fast and loose with the facts in order to generate interest or perhaps push a particular point of view. I submit to you however that Oliver Stone in his wildest flight of fancy will be hard pressed to equal the distorted view of history found in 'Inherit the Wind.' The question then becomes, why? Why twist the facts in such a way to make the agnostics appear kinder, brighter, or heroic than they really were? Why the dishonest attempt to lampoon Bible believing Christians?"[14]

Well, I think it's pretty obvious, isn't it? It's the first "deceptive" step into getting the American people "away" from the knowledge of "God" and where our "rights" and "freedoms" come from so that these "Aliens" I mean rich "Elites" can drag us back into "slavery" and "tyranny." And now, the utter hypocrisy is, this "lie" of Evolution and Humanism and Atheism is "all we get" in our Education System, that again, is not only contrary to the facts of Science and Logic and common-sense reason, but also the original beliefs, desires and wishes of our Founding Fathers for our country to ensure we never go back to "tyranny" and maintain our "God-given" "rights" and "freedoms." In fact, this deceptive hypocritical Humanistic stranglehold on our Educational System is so entrenched, that they not only refuse to have any other view of origins like the Biblical view, even though they said this was "unfair," but they literally demand "no intelligence" be allowed in our Education System either, as this man exposed!

Professor: *"Moving through history in an unguided and undesigned way the theory of evolution…"*

Ben Stein: *"Excuse me!"*

Professor: *"Yes, Ben."*

Ben Stein: *"How did life begin in the first place?"*

Professor: *"Mr. Stein, you have the same question every time."*

Ben Stein: *"Well you never answer."*

Professor: *"We've developed it. You know we have been through this so many times.*

Ben Stein: *"Could there have been an intelligent designer?"*

Ben is sitting outside the Deans Office and a kid sitting next to him on the bench asks why he is there. His answer is, *"I made a movie."*

Narrator: *"Join icon Ben Stein in this year's most controversial, documentary film."*

"If they value their careers, they should keep quiet about their intelligent design views."

"I was viewed as an intellectual terrorist."

"I have never been treated like this in my 30 years in academia."

"I lost my job."

"It's a funny thing that questions that aren't properly answered don't go away."

Ben Stein: *"How do we get from an inorganic world to the world of the cell?"*

"It might have started off on the backs of crystals."

Ben Stein: *"You have no idea how it started."*

"No."

Ben Stein: *"So intelligent designers believe that God is the design."*

"God is about as unlikely as fairies, angels, hobgoblins, etc."

"Science makes no use of the hypothesis of God."

"I mean it's essentially official policy of the National Academy of Science that religion and science will not be related."

Ben Stein: *"There are people out there that want to keep science in a little box where it can't possibly touch a higher power, cannot possibly touch God."*

Back to the Deans Office the kid says, *"Must be some movie."*

"Expelled, no intelligence allowed."[15]

     I.E., even though Science, logic, reason, facts, and common sense shows us that there really is a "Creator" Who gives us these "unalienable rights" and "freedoms" to protect us from the tyranny of "man." As you saw, our so-called Modern Education System refuses to "allow" that kind of common sense "intelligence." Even to the point, as you saw, they will actually "fire" teachers, or any staff member or faculty who has the audacity to "disagree" with their theory of evolution based on their own "scientific" discoveries! By the way, that's not "science." That's not "free thought" nor discussing where common sense "facts" simply lead you! That's what's going on here folks in our Education System! It's "indoctrination" not "education" into a "baseless" belief system designed to "brainwash" them against "God" and where our "freedoms" really come from, so they'll "mindlessly" go along with a man-centered nefarious agenda by these "Aliens" I mean rich "Elites" to "control" them around the whole planet with the "exclusion" of any "intelligence!" They literally want you to check your brain in at the door or you will be punished! Again, if you want even more proof on these deceptive "Elite" tactics and the "lies" they're promoting in our Education System, get the Ben Stein movie you saw there, "No Intelligence Allowed" or our own study again called, "The Witness of Creation." But again, as you can see, these kids today are not being told "how to think" but "what to think." No intelligence is allowed because in order for these "Aliens" I mean rich "Elites" to pull off their "mesmerizing of the minds of the masses" to fulfill their evil "nefarious agenda," they have to "reeducate" and "brainwash" the public away from Biblical principles of "freedom" like our Founding Fathers set up, and instead supplant them with the Humanist Atheist Evolutionary "lies" that get people to think mistakenly that our rights come from "man" not God.

This is why they hijacked the Education System, and this is why we now have a whole new generation of young people called Snowflakes who can't handle any differing opinions other than their own, which is really the opinions that the "Elites" want them to have. This is why they have to have "cry closets" until the Government "bans" "burns" "deletes" and "forbids" any "free thought" or "personal opinion" outside of what the "Elites say" should be the norm. These Snowflakes are now "brainwashed" into thinking that the Government is the one who will protect us and who should also dictate every aspect of our lives and then even "promote" it as a good thing for us. This "brainwashed" generation is simply the long awaited for fruit of the "Elites" hijacking of the Educational System to go along with their nefarious "Alien" global agenda that began many decades ago. In fact, speaking of "fruit," wonder of wonders, this hijacking of the Education System that used to be built on Biblical principles of freedom, and supplanting it with the lie of Evolution, Atheism and Humanism not only has produced a generation of "sheeples" who will go along with whatever the "Elites" want them to go along with, but it has also given rise to the "immoral" destructive society that God warned about that would come on the scene when you're living in the Last Days. As was mentioned earlier, after the Scopes Monkey Trial, the theory of Evolution began to be taught more and more in our Education System and then in the early 1960s, prayer and Bible reading were taken out. And the results have been disastrous! Let's take a look at the official behavioral statistics from the U.S. Census Bureau from about 1963 on, and you tell me if evolutionary teaching hasn't had a horrible "immoral" impact on our country like God warned would happen.

- Back in the 1950s, the average textbook only had two to three thousand words about evolution. But in 1963, it jumped up to 33,000 words. And it just so happened that 1963 is also when prayer and Bible reading was taken out of the American school system. What "moral" effect did this have on our country?
- Violent Crimes are up 995%.
- The inmate population has grown from about 200,000 to nearly 2 ½ million or about one out of every 99 adults. America now has more people per capita in prison than any nation on earth.
- Unmarried couples living together are up 725%.
- Illegitimate births are up 400%.
- The divorce rate is now 60%.
- A murder occurs every 23 minutes.
- A rape occurs every 6 minutes.
- An aggravated assault occurs every 48 seconds.
- An abortion takes place every 20 seconds.
- A burglary occurs every 8 seconds.
- Sexually transmitted diseases among 10-14 yr. olds are up 385%.
- Unwed pregnancies among 10-14 yr. olds are up 553%.
- Teenage suicide is up 200%.
- The national average of SAT scores has dropped 80 points.
- Across America, 25% of graduating seniors couldn't read their diploma.
- Every day 13 youth commit suicide, 16 are murdered, 1,000 become mothers, 100,000 bring guns to school, 2,200 drop out of school, 500 begin using drugs, 1,000 begin drinking alcohol, 3,500 are assaulted, 630 are robbed, and 80 are raped.
- More people have died in America in the last 30 years from murders and suicides then from all the wars in the history of the United States.

Why? Because this is the "fruit" of your "lies." If you "lie" to the kids and tell them that there is no God, and then "brainwash" them into thinking they came from an ape, then why are we surprised when they act like apes! What you believe determines how you behave and if you don't believe there is a God, you're going to act like it! You're going to act "ungodly!" Exactly what God warned about would come on the scene when you're living in the Last Days! This evil Humanist, Atheist, Evolutionary agenda has produced it! And now, people are looking around and wondering why things have gotten so "violent" and "wicked" and "immoral" in our country, including around the world, and it's simply because we've all been duped! We allowed our Education System to get "hijacked" by "Elite" Humanist Atheist Evolutionists who have kicked God out, as this video transcript shows.

Student: *"Dear God, why didn't you save the school children at Moses Lake, Washington; Bethel, Alaska; Pearl, Mississippi; West Paducah, Kentucky; Stamps, Arkansas; Jonesboro, Arkansas; Edinboro, Pennsylvania; Fayetteville, Tennessee; Springfield, Oregon; Richmond, Virginia; Littleton, Colorado; Tabor, Alberta Canada; Conyers, Georgia; Deming, New Mexico; Fort Gibson, Oklahoma; Santee, California; El Cajon, California; Blacksburg, Virginia? Sincerely, concerned student."*

Reply: *"Dear concerned student, I am not allowed in schools. Sincerely, God."*

Narrator: *"How did this get started? I think it started when Madelyn Murray O'Hare complained that she didn't want any prayer in our schools. And we said okay. Then someone said, you better not read the Bible in school. The Bible says, Thou shall not kill. Thou shall not steal. And Love your neighbor as yourself. And we said okay.*

Dr. Benjamin Spock said we shouldn't spank our children when they misbehave, because their personality would be warped, and you might damage their self-esteem. And we said, an expert should know what he is talking about. So, we won't spank them anymore. Then someone said, teachers and principals better not discipline our children when they misbehave. And the school administrator said no faculty member in the school better touch a student when they misbehave, because we don't want any bad publicity. And we surely don't want to be sued. And we accepted their reasoning.

Then someone said, let's let our daughters get abortions when they want, and they won't even have to tell their parents. And we said, that's a grand idea. Then some wise school board member said since boys will be boys and they are going to do it, anyway, let's give our boys all the condoms they want. So, they can have all the fun they desire. And we won't have to tell their parents that they got them at school. And we said, that is another great idea.

Then some of our top elected officials said, it doesn't matter what we do in private as long as we do our jobs. And we said, it doesn't matter what anyone, including the President, does in private as long as we have jobs, and the economy is good. And someone took that appreciation a step further, and published pictures of nude children and then went steps further by making them available on the internet. And we said, everyone is entitled to free speech.

And the entertainment industry said, let's make TV shows and movies that promote profanity, violence, and illicit sex and let's record music that encourages rape, drugs, murder, suicide and satanic themes. And we said, it's just entertainment. And it has no adverse effect. Nobody takes it seriously anyway. So, go right ahead.

Now we are asking ourselves why our children have no conscience, they don't know right from wrong and why it doesn't bother them to kill strangers, classmates, or even themselves. Undoubtedly, if we thought about it long and hard enough, we could figure it out. Surely it has a great deal to do with, we reap what we sow."[16]

In other words, you "lie" to people there is no God, they're going to start acting like it. But this "hijacking" of the Education System by these "Elite" Humanist Atheist Evolutionists, in order to "brainwash" people away from God and the knowledge of where our "rights" and "freedoms" come from, in order to trick us into going back into "slavery" and "tyranny" not only is producing the "fruit" of the "immoral" society that God warned about would come upon the scene when you're living in the Last Days. But it has also simultaneously produced the "scoffing" society that God also warned about would come on the scene when you're living in the last days! And for proof of that, let me share with you some quick stats on Atheism in America and you tell me if it's getting more and more popular by our rejection of God via this "lie" in our Education System called Humanism, Atheism and Evolution.

- The number of Americans with no religious affiliation has grown by 25% over the past five years.

- According to the U.S. Census Bureau, the number of Americans with "no religion" more than doubled between 1990 and 2008.

- A study conducted by the Barna Group discovered that nearly 60% of all Christians from 15 years of age to 29 years of age are no longer actively involved in any Church.

- It is being projected that the percentage of Americans attending church in 2050 will be about half of what it is today.

- According to LifeWay Research, 46% of all Americans never even think about whether they will go to heaven or not.

- Atheism ads are now appearing all across America. Hundreds of thousands of dollars are being spent on Billboards declaring "We Are Good without God," or "Don't Believe in God? Join the Club!" or "Praise Darwin! Evolve Beyond Belief" and "Christianity: Sadistic God; Useless Savior, 30,000+ Versions of "Truth" Promotes Hate, Calls it "Love" just to name a few.

- Atheists in Florida are scrubbing away a blessing with unholy water that Christians prayed for on a highway. They say it made them uncomfortable and that they were not going to tolerate bigotry.

- Various atheists across the country are hosting "Rapture Parties" because "If the Rapture indeed occurs, and Christians worldwide are transported to heaven, we know as atheists, we're not going. If it occurs, it's a good thing for us. We get the real estate and cheap cars, and we won't have to worry about separation of Church and state."

- There's a new website out there called (www.if-jesus-returns-kill-him-again.com) where atheist Darwin Bedford, the self-proclaimed "atheist messiah" states among other blasphemous things, "If that self-made "blank" Jesus returns again, I think we should hire some Jews to throw Him off the rooftop once again. Or maybe we could hire some of Mel Gibson's extras from "The Passion of the Christ" to punish Him first."

- And if that doesn't work, believe it or not, there is now a new web filter out there for the Internet called, "God Block" that blocks all religious content to protect kids from, "The often violent, sexual, and psychologically harmful material in many holy texts, and from being indoctrinated into any religion." And the reason why they say they developed this device for parents and schools was because there's been a resurgence of fundamentalist religion.

So, as you can see, we now have a "scoffing" society who's vehemently and in many cases violently against God, Jesus, the Bible and Christian teachings. And it's gotten so bad that they are now doing such "God-hating" things like what I'm about to show you. The first clip is a new "atheist" summer camp for kids, followed by a hairdryer "de-baptism" service by an atheist encouraging kids to "scoff" and "mock" at God, and then it goes downhill from there.

BBC News: *"Wake up! It may be a summer holiday but it's no time to lie in bed at Camp Quest. Twenty-four children ages 7 to 17 are attacking rational thinking and a scientific approach to the world. An array of summer camp activities. They also learn cooperation, tolerance, and empathy in an atmosphere free of religious propaganda. Most of the parents are atheists and they sent their children to learn ethical behavior separate from religion."*

Samantha Stein, Camp Quest Organizer: "I think people are possibly getting tired of the influence that religion has in society, possibly an unearned influence, and they are trying to come up with alternatives."

BBC: "With the sun setting, it was time for a little open-air education. Mainly it is encouraging children to think skeptically and not simply to be told what to think. They are likely to talk about religion and be told that religious beliefs can hinder morale behavior."

At an Atheist De-baptism Ceremony, a man dressed in blue is holding a blow dryer, holding it over the adults and kids as they are passing in front of him. He tells them not to worry, they are fine. They won't go to hell. In symbolism he is drying off the water of the water baptism that they previously had de-baptizing them.

CNN Larry King: *"We are going to run the unedited version. And be warned people, you may be offended."*

Kathy Griffin at the Emmys: *"Now a lot of people come up here and they thank Jesus for this award, I want you to know that no one had less to do with this award than Jesus. He didn't help me a bit. So, all I can say is, suck it Jesus. Jesus this award is my god now."*

KENS5 News: *"A program that has been turning heads and dropping jaws on the UTA campus Friday, is in the home stretch. Through tomorrow an atheist group is offering free pornography if people are willing to give up their Bibles. It made more than a few people angry, but the University said the group has not broken any rules and they should be allowed to exercise their rights."*

SML News: *"Now is it possible to have a church without religion? The Sunday Assembly believes it is. The organization says in its gatherings it extracts the good things out of religion without making God part of the package. We went to one of their get-togethers and we discovered a church with a vicar that he doesn't believe in the existence of God either. While many people may find the idea of a godless church strange, one*

Anglican Vicar has been preaching for over 40 years despite never believing in God."

Anglican Vicar: *"Atheism and religion don't have to be enemies."*

SML News*: "They come together regularly, and their numbers are growing. The Sunday assembly was started by Sanderson Jones and Peter Evans months ago. Now there are eight congregations worldwide, with the main branch in London."*

Sanderson Jones: *"We are a godless congregation that celebrates and are there for people that want to live better, help often."*

SML News: *"The idea of the assembly is to take what they call the best part of church and celebrate life, atheists and secular respective."*

CNN: *In today's Faces of Faith, we are talking about a new kind of church. A church without God. It's called Community Mission Chapel. It looks and sounds like a place of worship. There is a weekly service, there are offerings, uplifting songs, but here is the catch. It is missing one major detail. Its members don't believe in a higher power. This self-proclaimed atheist church is just one example of the growing religiously unaffiliated congregations popping up all over the world, and it's making some waves in the heart of the Bible Belt, Lake Charles, Louisiana."*[17]

And what did God say? There would be a massive rise of "scoffers" towards His existence and His soon coming judgment, when you're living in the Last Days. And right now, we live in a world, as you just saw, where we have atheist summer camps for kids, de-baptism services, celebrities openly mocking God with the audience laughing and applauding, pornography being given out instead of Bibles and so-called Atheist Churches spreading across the world, even in the Bible belt of America. Why? Because we allowed the Humanist, Atheist, Evolutionary "Elites" to hijack our Education System and now they have successfully "brainwashed" a whole generation into "scoffing" and "mocking" at God and the things of God. The "Aliens" I mean rich "Elites" have simply used the Schools around the whole planet to "mesmerize the minds of the masses" into going along with their evil nefarious agenda.

Again, anyone starting to see a pattern here with all these different forms of media? All of them are being used to "control the narrative" and to "manipulate the minds" of the masses for these "Elites" "own" nefarious purposes. This Media Manipulation has not only led to an "increase of wickedness" in the Last Days, like the Bible warned about, but it has also simultaneously produced a global society of "scoffers" towards the existence of God and His soon coming judgment! Total "manipulation" of the "minds of the masses," across the whole planet, 24 hours a day, 7 days a week, just by simply "owning" and "controlling" the media called Newspapers, Radio and Music, Books, and now the whole Education System. The whole planet really is being bombarded by all five of these forms of Global Media and there's no escaping it!

Unfortunately, it's about to get even worse. These "Aliens" I mean rich "Elitists" are not merely content with just owning all the Newspapers, and Radio Broadcasts and Music Industry, and Books and the Education Systems around the whole planet to "control the minds of the masses." In their insatiable lust for "power" and "domination" they really are literally out there amassing "all forms of media" in order to get the job done. Including Television. It too is being used as a powerful Global Media tool to "subliminally" get us to "think," "act," "believe," or "behave," the way these "Elites" want us to. This will be the topic of our next section. And as we will soon see, the Media of Television is also telling us "to not think for ourselves," "to obey only what they want us to obey," "to never question them," and of course, "to scoff" specifically at the Bible, God, Jesus, and His Soon Coming judgment! Where have I heard that before? I know it's hard to believe, but once again, "keep those glasses on," and let's begin this next journey of exposing how this next form of "media" called Television is also "mesmerizing the minds of the masses."

Chapter Six

The Manipulation of Television

The **4th way** the global media is "controlling the narrative" to "control our minds" is with **Television**. You see, Newspapers, Radio, Music, Books and the whole Education System are not the only Subliminal Manipulative Technologies that are being used on us on a global basis by a small group of "Aliens" I mean rich "Elites" around the world. Believe it or not, the whole Television Industry is as well. It too is out there on a massive scale "swaying our minds" subconsciously "telling" us to "buy" and "consume" and even "instruct" our minds "to go back to sleep," "never question authority," "obey" and "believe" whatever these rich "Elitists" "want us to believe" including the need to "scoff" at the things of God. In fact, it's one of their biggest and most powerful tools to get their "manipulative" job done as we saw in the first two sections of our study dealing with the ongoing Subliminal Technology that's still being used on us today. However, believe it or not, that's the tip of the iceberg when it comes to Television's "subliminal" influence. So again, let's "keep those glasses on" and "see" how this next Media called Television really is being used to "tell a vision" alright, for these "Aliens" I mean rich "Elites" in order to further "manipulate the minds of the masses" on a huge global scale. And again, just like we saw with the other Manipulative Technologies, so it is with this "manipulation" called Television. Its subversive abilities have been going on for quite some time now, whether people realized it or not. You see, you thought TV was just for "entertainment" or even possibly "education" purposes. Not at all. It's for "manipulation" and "indoctrination" on a scale most people have a hard time believing. In fact, let's see how this "subliminal" technology began.

The Evolution of Television

Narrator: *"In 100 years, the TV has taken many shapes and sizes. Here's the evolution of the television from the 1920s to today. The 1920s gave us the mechanical television. The first model had a small display on the right in a huge cabinet. These first TVs were very simple in comparison to our technology today. They implemented peculiar shapes such as the Octagon television. Although the technology was impressive, the video quality was not. Facial features were not recognizable unless makeup was worn in a specific way.*

In the 1930s we saw more refined televisions with better designs and resolution. This decade saw a giant leap in video quality. From 100 scan lines at the start of the decade to 405. The 1940s brought us even higher resolutions including the NTSC standard of 480 lines of resolution and better audio. In the 1940s it was difficult to produce CRT screens larger than 12 inches, so big screen TVs implemented projection techniques.

The 1950s brought us the short-lived porthole televisions but most importantly it brought us the NTSC color standard, but not many color televisions were sold until the following decade. The sales of color TVs boomed in the 60s. They were now more affordable and the colors more vivid.

The 1970s brought TV designs for every taste and need. There was the upgradable television with easily accessible circuit boards. The futuristic televisions which implemented curvy design patterns. The colorful TVs that looked more like toys and the portable combos which normally included radios and cassette decks.

The 1980s was the end of TV as furniture era. These were replaced with color projection TVs with larger screens and minimalist cabinets. The space command was one of those color projection TVs. The 1980s gave way to the first LCD TVs. They were tiny and pixelated, but it was a huge step forward. These portable TVs were a lot more affordable and included additional functions. This decade was all about Sony's Trinitron technology. As Sony's patent had run out, all competitors were free to use the technology. Combos became really popular in the 1990s. They included FM radio and VCR. In the 1990s Casio continued to improve their portable LCD TVs.

A preview of one of the biggest techs of the following decade was introduced at the end of the 1990s. The flat HD TV priced around $7,000.00. Throughout the 2000s CRT TVs were still very popular, as they were inexpensive and could access HD channels using a converter box. During this decade LCD tech was finally able to compete with plasma giving us LCD HD TVs.

Then LED TVs arrived which were superior to both. A few brands experimented with ambient lighting to make the TVs more immersive. Through the decade HDTVs became pretty smart. They were able to display content from your computer as well as to connect to different services on the internet.

We started in 2010 with the best combination of tech money could buy, 4K, LED and 3D which had all just been introduced. Curved screens became the premium feature for a few years. It was meant to reduce glare and improve immersion by taking advantage of our peripheral vision. This works well for computer monitors but it is barely noticeable on TVs.

Something truly impressive was the introduction of wallpaper TVs. These are almost as thin as a credit card. The latest feature is ambient mode which lets you match your TV to your décor. It allows you to display your TV as art or to match your wallpaper.

2020 promises many new technologies such as the rollable TV. It is completely hidden when it's off and when it's on it grows to the size of the content or application. This is great for calendar and music apps and avoids the black bars and extra-wide movies. This is another discreet TV technology, but it will likely be used more for decoration and art than for entertainment. The double-sided TV will be great for gaming and for commercial applications. Tech is changing really fast and so are our TVs. What do you think is next for the evolution of television?"[1]

Well, believe it or not, it's going to skip the screen altogether and go straight into your brain! And if you don't believe me, hold on a second and I'll show you that reality is "already" here as well. But as you can see, with the advances of Television over the past 100 years or so, it has not only become a major Media staple to "mold our minds" and allow us to "receive information" in general. But just like Newspapers, Radio, Music, Books, and the Education System, once this "manipulative" Technology called Television sprang into existence, it too began to spread across the whole planet. That is why today, we have an estimated 28,000 Television stations around the world via cable and satellite signals allowing the broadcasting of all kinds of "manipulative messages" into the "minds of the masses" on a global scale. In fact, even in recent years, Television is now morphing into APP based streaming technology via the Worldwide Internet where you can "customize" your own "manipulative inputs" and get just about any kind of TV Channel you could ever think of, or want, on just about any kind of topic anywhere in the world!

Another fantastic leap forward of Manipulative Media Technology if you want to connect the world and "mesmerize the minds of the masses" for some nefarious purpose! I wonder who and what that might be? It rhymes with the "Elites" again for those of you wondering! And speaking of "Aliens" I mean rich "Elites," for those of you who are still wondering about my earlier statement about how they are going to skip the whole Television screen altogether and go straight into your brain, let's go ahead and expose that, shall we? Believe it or not, they have not only figured out a way to "mesmerize the minds of the masses" when we are awake with all the various forms of Media with Newspapers, Radio, Music, Books, the Education System, and now even TV, but they have also figured out a way to get that "manipulative broadcast" straight into your mind with a microchip implant, which will then allow them to "dictate" your dreams while you sleep. Don't believe me?

Julia Sieger, France 24 Reports: *"Can we trust our own brains and what can our dreams teach us? Here is Moran Cerf, a neuroscientist from Northwestern University. Thank you for being with us today. Is it possible to hack a human brain and is it as easy as hacking into a computer?"*

Moran Cerf, Neuroscience Professor, Northwestern University: *"It's actually possibly easier. The idea is we are now starting to understand how you can influence a person to change their mind and behave a certain way. Many companies are interested in that because this tackles the very idea of free will. Where Amazon can not only know what you want, but they can actually start offering something that you didn't know that you want. Because they know you better than yourself. That is where they are heading into your world where your brain is actually in the outside world behind your back."*

Julia Sieger: *"Now you have also studied dreams quite a bit. Can deciphering their meaning be a key to our brain?"*

Moran Cerf: *"Yes, they have been fascinated with peoples dreams for a millennia. We always thought that they meant something. If you don't believe me, you can just try to tell your husband or your boyfriend in the morning that you dreamt of your ex-boyfriend and see how they respond. They think things mean something, whether they do, or they don't. Now for the first time in history, we can access it so we can understand it.*

Up until now with the days of Freud and all of those, they had to ask for you to tell us your dream for them and believe this is the story but what we are learning right now is that we can actually get access to your dreams using neuroscience, while you are sleeping, to understand what they are telling us. They can tell us something about who you are, what you want, what you think about the future, what you want to do next, how you think about all kinds of possibilities. With this we can get to understand you better, it can actually control the world around you better, and we can even start to manipulate your dreams by creating dreams for you. So, you can go to sleep and dream what I set up for you. Maybe a film maker like Stephen Spielberg can make dreams for people and so on and at the very least you can go to sleep after a nice evening and continue the evening in your sleep. Companies like Netflix, Hulu, YouTube or Tech24 can put content in, generally it's like virtual reality that our brain creates for us."[2]

Or should we say, one that the "Elites" create for us? And you thought I was kidding when I said they've skipped the whole Television screen altogether and went straight into the brain? Yeah, I wish I was making it up too, but as you saw, it's unfortunately and shockingly true. But talk about total "manipulation" on a level never before "dreamed" of, pun intended. Imagine tricking the whole planet into getting a microchip implant in your head to do all this, to watch TV in your brain and get "streamed dreams" let alone any other kind of "media input" from these "Elites." I mean, what are they going to call this new "manipulative" service, "Dreamflix?" But if you haven't already caught on, this is all starting to sound like another movie out there called, "The Matrix." Do you remember the scene that exposed the "fake generated reality" that the whole planet was under, produced by a ruling "Elite," showing how people were being used for a nefarious purpose? Let's look at that again.

A Clip from the Matrix

The scene opens with a large bag of liquid. There are gurgling sounds coming from the bag and there are black wires running through the liquid. Something inside is moving around when suddenly a hand reaches out. As it extends further a form of a human bursts out of the lining of the bag. A full-grown human is gasping for breath. He immediately starts grabbing hold of the wires and disconnects them from his body. He seems to be having trouble breathing, when he reaches for the cord that is connecting his head to the machine. He starts to pull it loose when he sees that there are several other pods just like his with people in them as well. But then as he stands to look over the edge of his pod, he looks up and down and realizes there are thousands, maybe millions of these pods all lined up in layers on towers. Suddenly a voice calls out, "Welcome to the real world."[3]

Yeah, the real world that only those who have the right set of "glasses on" or in this case, who take the "red pill" and can "see" "manipulative illusion" or the "fake world" that these "Aliens" I mean rich "Elites" have created for us to "enslave" and "control" us for their nefarious purposes. As you can see, with the Technology, including Television, they're already out there promoting a chip in the brain to create a "matrix" type reality for us. Through Television, they're not only getting us to go along with it, but again, to actually get us to think it's a good thing for us. Anyone starting to see a pattern again here? And as far as people actually getting a microchip in their brain so an "Elite" can "mesmerize the minds of the masses" on a global scale for their nefarious evil purposes, we already know that's exactly what's going to happen to the planet in the Last Days. Here's yet another warning from God it's getting close!

Revelation 13:16-17 "He also forced everyone, small and great, rich and poor, free and slave, to receive a mark on his right hand or on his forehead, so that no one could buy or sell unless he had the mark, which is the name of the beast or the number of his name."

Good thing we don't see any signs of people getting some sort of "mark" in their "hands" or "heads" to "buy and sell" including "buying" Television "streaming" or "dreaming" content. As crazy and wild as that sounds, it's already here, as you just saw! It's almost like somebody's following a script or something and God knows the future and He's trying to warn us of it! But as you can see, it's common sense. Television has become a very important tool to share "information," "educate," or even "mold" the minds of people either for good or bad. Therefore, once again, whoever controls the Media of Television, in all its various forms, has yet another powerful tool to "manipulate the minds of the masses" around the planet. And believe it or not, that "control" of the Media of Television on a global scale is already here! Just like with Newspapers, Radio, Music, Books, the Education System, so even Television is now in the hands of just a few "Aliens" I mean rich "Elites." In fact, they're not even hiding it either. Are you ready for this? Talk about a pattern. They call themselves, wait for it, "The Big Five." Just like the Books and Textbooks, isn't that wild? But here they are, owning "all" the Television outlets.

ESPN, Vice, and the History channel, what do these three media outlets have in common? Despite their different target audiences, they are all owned by the same company, Disney. What about these three. Syfy, Focus Features, and NBC? Same story, they are all owned by Comcast. If you were to pick a bunch of media operations from a hat, odds are that a vast majority of them are owned and operated by one of five mega corporations.

Since the early 1980s the number of media companies controlling the bulk of US media have run from 50 to just 5. We have gone from a media landscape operated by a reasonable number of controlling interests to a reality dominated by mere monopolies, just barely sneaking under the enforcement of anti-trust laws. In this episode we're going to talk about the consolidation of American media in the hands of a few ultra-powerful companies. And what that means in a time when the very wealthy have outside influence over public policy.

Before we begin it's worth noting that corporate deals, acquisitions and mergers happen all the time, often with little to no coverage. It's quite possible that by the time you see this, some of these data points will have changed. But unless something truly cataclysmic happens, the vast majority of US media will still be under the thumb of one or more of the five super corporations we're about to discuss. So, without further ado let's meet the big five.

Comcast, Disney, National Amusements, News Corp. and AT&T, taking the spot from Time Warner which it acquired in 2018 for $109 billion dollars. You probably guessed that Disney was on the list since they've had some seriously high-profile acquisitions in recent years, most notably Marvel and Lucasfilm. But I'd be surprised if you've ever heard of National Amusements or News Corp. These incredibly bland names and their lack of public-facing operations are intentional. Why draw attention to yourself as one of a tiny number of corporate behemoths when people only care about a select few of your properties.

Whether you are watching CBS News, scrolling through GameSpot or reading a book published by Simon and Schuster, you're padding the enormous wallet of National Amusements. And you'd never suspect that these three very different operations are run by the same company. The same goes for News Corp. They own National Geographic, Fox News and Harper Collins.

Altogether the big five are worth over $400 billion dollars controlling something like 90 percent of all US media, including news networks, Hollywood movie studios and print publications and they reach nearly 100 percent of all US households. And, fun fact, much of the remaining 10 percent is owned by only slightly less giant multi-billion-dollar corporations, like Sinclair which as you may remember got itself into hot water by blasting out a propaganda broadcast over hundreds of local US news channels.

So how exactly are they allowed to do this? Surely having just five companies dominate the nation's media is considered monopolizing, right? Well, no, but just barely. Let's look at what it takes to be considered a monopoly. In order to be considered a pure monopoly, a single company has to have complete control over a market containing a goods or services with no close substitutes. So, there you go, since there are five large corporations that share the US media market, none of them qualify as a monopoly. But you don't have to be classified as a pure monopoly in order to wield monopolistic power.

With such a massive concentration of wealth and power in the hands of so few it all but ensures that smaller operations never have the chance to *succeed unless you count the very American concept of success which is building a company just large enough to get bought out by one of the big players. What we see in the US media landscape is more of an oligopoly. Complete control spread across just a handful of powerful groups. There's so much interplay and deal-making within these five mega-corporations that in effect they are a monopoly in every sense but the legal one.*

For example, it is not uncommon to see one of the big five hold a massive stake in a property owned by another of the big five. But if there were so many more media corporations as recently as the 1980s, how did we get to the dystopian all-powerful corporate landscape we see today? To make a long story short it all boils down to the passing of the 1996 Telecommunications Act. This piece of legislation was supposedly intended to deregulate the increasingly tangled broadcast and telecommunications markets allowing anyone to enter and compete in the industry. Whatever the intended result was, the actual outcome was simply the rapid consolidation of power in the hands of fewer and fewer massive corporations.

In the 2003 edition of Howard Zinn's "A People's History of the United States" he notes, "The Telecommunications Act of 1996 enabled the handful of corporations dominating the airwaves to expand their power further. Mergers enabled tighter control of information." He was right and it only got worse. The decade before the Telecommunications Act, 50 companies controlled the majority of the media landscape.

By 1992 that number had fallen by 50 percent. After the passage of the new TV and Broadcast Legislation, the number quickly shrank even further to just 6 in the year 2000, and that's roughly where it stayed to this day. Not because these giant companies wouldn't love to own more of the market, but because they physically can't without triggering anti-trust lawsuits. To put into perspective just how much of the market the big five control let's have a look at some of the media operations they own.

We'll start with News Corp. Rupert Murdoch's empire owns Fox including all of its branches like Fox Sports and 20th Century Fox. It owns FX, GQ,

The Wall Street Journal, Sky News, Harper Collins Publishing, The New York Post, National Geographic, Zondervan, Market Watch, and countless others.

National Amusements owns CBS and its branches. It owns Paramount, Nickelodeon, MTV, BET, GameSpot, VH-1, Comedy Central, The Smithsonian Channel, Spike, Showtime, Simon and Schuster, Game FAQs, Cnet, and Viacom, once a major player in the media world itself before being consumed.

AT&T, the newest member of the big five after acquiring the massive Time Warner, controls CNN, HBO, Cartoon Network, Warner Brothers, DC Comics, TBS, True TV, Cinemax, TNT, Adult Swim, part of Hulu, Turner Classic Movies, Time Magazine, Rocksteady Games and Time Warner Cable, to name just a few.

Comcast, which has a well-deserved reputation as a thoroughly evil corporation, owns NBC, MSNBC, USA Network, Syfy, Fandango, Universal Pictures, Focus Features, Working Title Films, Rotten Tomatoes, Bravo, Oxygen, Big Idea, part of Hulu, MLB Network, NHL Network and dozens of internet ventures.

And finally, everyone's favorite family-friendly corporate overlord, Disney. Disney owns Disney, ABC Pixar, DreamWorks, ESPN, Lifetime, The History Channel, Marvel, Lucasfilm, Hollywood Records, Touchstone Pictures, Vice, plus a giant swath of the comic book industry thanks to their acquisition of Marvel. And of course, these are only a small sample of the media operations owned by the big five. Odds are, if you can think of a network you watch, you'll find it's owned by one of them.

The one main exception you may have noticed is Netflix. Netflix, while not outright owned by any of the big five is owned in part by a number of large interests. Some of them very shady, like the well-known Blackrock, the world's largest shadow bank. So really, no matter what your network preferences you're only being offered the illusion of choice. In reality, almost everything we watch or read whether on TV, online, in theaters, or *on the pages of a book, newspaper, or magazine is just a tentacle of the enormous kraken that is the corporate media landscape.*

No matter what perspective, these outlets offer remember that they're all owned by the ultra-wealthy business interest and they have their own agenda. The big five own all the major news networks and the messages they put out are designed to reinforce the status quo. Peddling minor aesthetic disagreements and diverting attention away from serious societal problems. This is simply the natural conclusion of the hyper-capitalist system we have in the United States. Powerful companies will grow larger and larger consuming smaller companies that can't compete, acquiring more and more properties until they've reached the very limit of what could be considered legal, thereby dodging anti-trust laws and maintaining the maximum amount of power, profit and cultural significance. This is only one area of the American market.

Every other aspect of American life is becoming similarly consolidated from pharmaceuticals to energy to manufacturing. The threat of the monopolization of every area of modern life is real, and its consequences could be disastrous and far-reaching, except for those pulling the strings at the very top.[4]

Who happen to be the "Aliens" I mean rich "Elites" we've been talking about who have literally created a total "Media Matrix" for us around the planet, whether we realized it or not, or want it or not, for their own nefarious purposes! And did you see, these same "Big Five" also own many of the other Global Media Outlets we've already covered, including Newspapers, Magazines, Radio, Music and Books, just to name a few? Do you think that's by chance? Not on your life! In fact, if you still don't think these "Elites" are using their global ownership of virtually all Media outlets, including Television, for their own evil "manipulative" purposes to regurgitate a "preprogrammed" "pre-scripted narrative" for "us" to digest from "them," let's watch that "Sinclair" broadcast he was talking about. You know, the one that let the "cat out of the bag" that showed how we are not really getting the "real" news and "real" facts from these News

Outlets, but rather a "preprogrammed" "pre-scripted" "managed news broadcast." The proof is shown here.

FOX29 News Reports: *"Hi! I'm Fox and Antonio's Jessica Headley." "And I'm Ryan Wolf."*

As you are listening to what the first newscasters are saying you hear voices in the background. Several Newscasters are speaking over each other with the same message.

KATU2 ABC Reports: *"Our greatest responsibility is..."*

NBC 2 News: *"is to serve our Treasure Valley communities the..."*

4 News: *"the El Paso Las Cruces communities..."*

"Eastern Iowa communities...."

"Mid-Michigan communities..."

CBS 4 News: *"We are extremely proud of the quality, balanced journalism that CBS 4 News produces but...."*

Now there are 9 different news stations that on the screen stating the same message. All talking over each other and saying the same thing.

Eyewitness News: *"Plaguing our country. The sharing of biased and false news has become all too common on social media, more alarming some media outlets that publish the same fake stories without checking facts first."*

While this newscaster is speaking it is also being said on several other stations at the same time. And the same words.

"The sharing of biased and false news has become common...."

"Unfortunately, some media push their own personal bias to control exactly what you hear, and this is extremely dangerous to our democracy."

Fox 29: *"This is extremely dangerous to our democracy."*

KATU 2 ABC News: *"This is extremely dangerous to our democracy."*

KATV 7 News: *"This is extremely dangerous to our democracy."*

Autria Godfrey, ABC 7 News: *"This is extremely dangerous to our democracy."*

Local 12: *"This is extremely dangerous to our democracy."*

NBC 6: *"This is extremely dangerous to our democracy."*

ABC 13: *"This is extremely dangerous to our democracy."*

Fox 11: *"This is extremely dangerous to our democracy."*

NBC 4: *"This is extremely dangerous to our democracy."*

As you can see the same message is being repeated over and over on all the news channels.[5]

Yeah, I would agree! Can you believe that? All "pre-scripted programming." And you wonder "why" they call these broadcasts for us to watch "programs." I think it's obvious. They exist to "program" our minds and "mesmerize the masses." And I know all this is a tough cookie to swallow, but as you can "see," if you "keep your glasses on" that is, we only "hear" and "see" what these "Elite" owned Corporations including those who own Television and News stations all over the world "want" us to "see" and "hear." It's all fake! It's totally generated! It's simply a real live "Media Matrix" created for our brains to be "manipulated" for their own nefarious purposes! Including the "silencing" of any "dissenting opinion" by people exercising their God-given "rights" to "freedom of speech." That truth came out in the last election cycle when it was shown that these "Big Five" owned Television outlets were "refusing" any "narrative" than what they wanted us to "see" or "hear" even from so-called "conservative" news outlets. Remember this censorship?

Fox News cut away from a new conference held by Kayleigh McEnany. The host explained there was no evidence to back up her "explosive charge" about election fraud.

Kayleigh McEnany, White House press secretary: *"This election is not over, far from it."*

Fox News: *"This is Kayleigh McEnany right now spelling out the Republican committees plan to fight what has been going on in some states notably Pennsylvania with the vote count."*

Kayleigh McEnany: *"We want every legal vote to be counted, and we want every illegal vote…."*

Fox News: *"We have to be very clear that she is charging the other side of welcoming fraud and of making illegal voting and unless she has more details to back that up I can't in good countenance keep showing you this. I want to make sure that maybe they do have something to back that up but that is an explosive charge to make, that the other side is effectively rigging and cheating. If she does bring proof of that, we will of course bring you that."*

American Agenda: *"Thank you Mike Lindell so much for joining us. So, what happened to your twitter account and the company page?*

Mike Lindell: *"Well, mine was taken down because we have all the election fraud with the voting machines, we have 100 percent proof and then when they took it down…"*

As he is trying to answer the question he is talked over by the interviewer and he can't complete the interview.

American Agenda, second interviewer: *"Mike, Mike, Mike, you're talking about machines that Newsmax has not been able to verify any of those kinds of allegations. We just want people to know that there is nothing of substance there. Let me read you something...."*

Meanwhile, Mike is trying to talk but they have turned down the volume on his microphone so you can't make out what he is saying, still trying to answer the question.

American Agenda: *"While there were some clear evidences of some cases of vote fraud and irregularities the election results in every state were certified and Newsmax accepts the results as legal and final and the courts also supported that view. So, we wanted to talk to you about cancel culture if you will. We don't want to relitigate the point that you are making. We understand where you are. So, let me ask you this. Do you think this should be temporary? It appears to be permanent, could you make the argument that it is temporary?"*

Mike Lindell: *"What?"* He could not hear the questions due to the interviewer talking over him as he was answering the first question.

American Agenda: *"Do you think this could be a temporary banning rather than permanent?"*

Mike Lindell: *"They want it to be permanent because you know why? Because I am revealing all the evidence on Friday of all the election fraud of these machines. And I'm sorry...."*

Again, he is interrupted so he cannot answer the question.

American Agenda: *"Can I ask our producers if we can get out of here, please, I don't want to have to keep going over this? We have not been* *about to verify any of those allegations... "* He gets up and leaves the interview.

American Agenda, First Interviewer: *"Mike, Mike, let's talk a little bit about...."*[6]

But you don't get to talk about what you want to talk about, as you saw, they'll either cut the feed, or walk off the set, after endless interruptions and changing the subject! Whatever happened to "free speech?" Even if the guy's wrong, let them finish will you! What's the big deal? Are you trying to "control" the "narrative" of what people "receive" from Television outlets to the point where you actually "cut the feed" and "silence" any dissenting view? That's not news reporting, that's news determination! It's indoctrination again folks! That's all it is, just like the other forms of media! It's all the same exact pattern! Even on so-called "conservative" news outlets! How much more proof do we need to "see" that these "Aliens" I mean rich "Elites" really are "controlling" all media including the News and Television stations, and we're only getting what these "Big Five" corporations want us to have, that is, their "preprogrammed narrative" that promotes their own nefarious agenda, not the truth! I'll say it again it's a Media Matrix created for us. We live in a world of "illusion" and "manipulation."

And as bad as that is, it's about to get even worse! As gut-wrenching as what we've seen so far concerning the "manipulation" of Television by these "Aliens" I mean rich "Elites, believe it or not, these same "Big Five" entities are not only "controlling" all the Television content around the whole world as you just saw, but they are actually using it to "subliminally manipulate" us with a format that's purposely "addictive" so we never "cut the feed" off from their daily "programming" of our minds! If you don't believe me, here's some of that proof. Even as far back as former President Calvin Coolidge, he admitted concerning the advertising industry, which is a big part of the Television industry:

"It is the most potent influence in adapting and changing the habits and modes of life, affecting what we eat, what we wear, and the work and play of a whole nation."

Furthermore, a New York Times Magazine article reported way back in 1966 that:

"TV is not an art form or a cultural channel; it is an advertising medium. It seems a bit churlish and un-American of people who watch television to complain that their shows are lousy. They are not supposed to be any good. They are supposed to make money."

You know, for the "Elites" as well as to be used to "brainwash" us with their "pre-packaged programming" and "controlled narrative" to get us to go along with their nefarious agenda. But again, it's not just there for the "watching" or "viewing." It's presented to us in a technological manner that's very "addicting." This is why we just can't shut it off! Don't believe me? Let's begin to expose that reality. First of all, do you ever wonder why Television and Movies in general are becoming more and more vulgar and obscene every single year? Well, it's merely the logical outcome of a society that's become what I call "Entertainment Junkies." This is what the "Aliens" I mean rich "Elites" have created us to be.

You see, a junkie needs an ever-increasing amount of drugs or stimuli in order to get their daily "fix." This is because they eventually build up a tolerance level and so if they want to get that same "high" then they must increase the stimuli. And so, it is with Television. These "Elites" know that they've already created a society of "Entertainment Junkies" and in order to "keep us glued to the screen" and not shut off the daily "programming" of our minds, they have to provide us with an ever-increasing number of "stimuli" from Television's both audio and visual inputs. This is why they are dishing out more and more vulgarity and obscenity to "shock" us into an emotional "high" or response in order to give us that entertainment "fix." Television is simply a global "drug" problem. And so just like a junkie who becomes immune to the effects of drugs over time, we too, the viewers have become "desensitized" to the "shocking" effects and things on TV and so consequently we need more and more "stimuli" in "bigger" quantities with even more "shocking" content if we're going to be "entertained" and "glued" to the screen. Now, as conspiratorial and wild as that sounds, observe some of the following facts from these Media experts.

A junkie is "consumed" with his "fix."

"Daily home usage of Television: 6 hours 59 minutes per day."

"With TV sets turned on in the inner city for eleven hours a day and multiplying satellite, cable, and broadcast channels, television has become the closest and the most constant companion for American children. It has become the nation's mom and pop, storyteller, babysitter, preacher, and teacher. Our children watch an astonishing 5,000 hours by the first grade and 19,000 hours by the end of high school—more time than they spend in class."

A junkie becomes "addicted" to his "fix."

"Once people are exposed to the spectacle of blood and sex, they want more and more as they become hardened to the titillation of the last violent or sexual act they see. Just as a drug addict who becomes less and less responsive to a drug keeps looking for the initial 'ideal' rush, so those who are addicted to the sex and violence in films seek increasing doses of sex and violence to appease their lust."

"Watching fighting or other violence can make the mind believe that it is about to engage in life-threatening activity. The body will often respond by releasing adrenal epinephrine into the bloodstream, giving the viewer an adrenal rush without the threat of actual violence. Watching sexual activity and nudity makes the mind think that the person is about to mate so the body releases raging hormones that can often cause an addictive adrenal rush without the psychological burdens attendant to most human relationships. The physiological phenomena will engage and attract the

viewer, often causing him or her to want more and more exposure to the stimuli that cause the artificial physical elation."

A junkie will "destroy" his mind for his "fix."

"Movies are one of the bad habits that corrupted our century. Of their many sins, I offer as the worst their effect on the intellectual side of the nation. It is chiefly from that viewpoint that I write of them—as an eruption of trash that has lamed the American mind and retarded Americans from becoming a cultured people."

"And because of television's insidious 'flicker,' (every four seconds, on the average, the picture changes) television does not promote long-term attention. Lastly, because the action shifts constantly and capriciously backward, forward, and laterally in time…television does not promote logical sequential thinking."

A junkie will eventually "kill himself" with his "fix."

"The point of this litany of problems with respect to different media and arts is simply to point out that there is no place to hide. The media and the arts are pervasive in our society. Americans are in the midst of entertaining themselves to death. Either denial or license will only allow the problems to continue to grow out of control."

"We often forget that there is a war raging around us. It is a war being waged inside our minds, a spiritual war for our souls. The adversary is using every possible tactic to control our minds: materialism, secularism, humanism, and all the other isms that conflict with Christianity. He is using the most effective weapons to win: the power of the mass media of entertainment. With the corrupted movies and television programs of our age, the adversary is fueling our sinful propensity to lust and hooking us on our desires. Once hooked, he drags us down to hell."

And this is precisely why, once you start to watch it, you just can't seem to shut it off! The "Elites" have got you "hooked" on their TV "programming."

Television: *"Look at me! Look at me! Look at me! Look at me! No, don't look over there. There's nothing over there! Look at me! Look at me! Look at me! Are you looking at me? Is everybody looking at me? Do I have everyone's attention? Don't get the wrong idea. I'm not trying to take over your life. You need what? What do you need? You need what? You need to go to the bathroom?? Fine, get up, go to the bathroom, come back, look at me! You need what? You need to get something to eat? Fine!!! Get up, go to the kitchen, get something to eat. Come back and look at me! You need to what, what, what? Get some sleep? Fine!!! Get up, go to bed, get some sleep, come back and look at me! So, we have an agreement? You will do what you absolutely have to do and when you are done, you will come back and look at me.*

Don't worry about your schedule. I am here for you! I am here for you 24 hours a day, 7 days a week, I am here for you. I am here for you! You need me! I am here fair and foul, thick and thin. I am here for you! I am here for you! People try and tell you I am bad! You tell them I am here for you, 24 hours a day, fair and foul, thick and thin, I am here for you! I am here for you!

People try and tell you I am bad. Do you know what it sounds like to me? Sour grapes. NO, NO, NO, don't look over there. There is nothing over there, look at me, look at me, look at me, I've got stuff you wouldn't believe. Danger, sex, action, death, thrills, comedy, all here, all in the next 8 minutes. You can believe it! You can't believe it! It's unbelievable! You can't believe it because it's unbelievable. It's a miracle, just keep looking at me! Just keep looking at me! Look at me, look at me, look at me, look at me![7]

Yeah, whatever you do, don't look away, look at the TV and keep looking at it, and the "Elites" will have successfully "brainwashed" you into not only "buying" things to fill their pockets with your cash, but to "program" you and "reprogram" you into "believing" "behaving" "acting" and "thinking" whatever they want us to. This is what Television is really all about. Then to make matters worse, this Television Technology not only is being used by these "Elites" as the biggest global "brainwashing" tool ever invented in the history of mankind, to go along with their nefarious agenda, but it also, believe it or not, puts the viewer into a "highly" suggestible even "hypnotic" state in order to make sure we "ingest" their daily "programming" straight into our brains without any resistance! Don't believe me? Check this fact out.

A junkie is "hypnotized" with his "fix."

"At the same time as television inhibits cognitive growth, research shows that children 'habituate' to repetitive light-stimuli (flickering light, dot patterns, limited eye movement). When habitation occurs, then the brain decides that there is nothing of interest going on—at least nothing that can be done about—and virtually quits processing information. In particular, the left brain 'common internegative area' goes into a kind of holding pattern, and television viewing reaches the level of somnambulism, similar to being hypnotized."

There you have it! Secular researchers admit it. There's no conspiracy theory here. TV is simply not only "brainwashing" us with its daily "programming" in an "addictive" manner so we can't shut the daily "programming" off, but it's also putting us into a "hypnotic" state to make sure we "believe" and "obey" the messages by purposely "embedding" it into our brains without any resistance! But if you need more proof, here you go.

Little Light: *"If you ever experienced a mind fog after watching television, you are not alone. The brain has four modes that it operates in. Four brainwave patterns. Delta is when you are deep asleep, Theta is when you are in light sleep. Alpha is when awake but relaxed, it's the mode of thinking that you are in when you're in the most heightened state of suggestibility. And then there is Beta, the highest functioning mode like* *when you are reading a book, or you are having a very stimulating conversation.*

In 1969 Herbert Krugman did a study where he found that in less than 60 seconds the human brain switched from beta brainwave patterns, those exhibited during logical thought to alpha brainwave patterns. Alphas not a bad brainwave pattern to be in but when you're in that brainwave pattern a lot of the time and if you look at the averages of how much people are watching television, this just leaves the mind unfocused and unable to concentrate. Researchers have said that watching television is similar to staring at a blank wall for several hours.

Robert Kubey, Director – Center Media Studies, Rutgers University, did a study which appeared in Scientific America in 2002.

Robert Kubey: *"Looking at how people experienced television in the flow of their everyday life and we used a beeper technology where we would beep people at intervals and they would write down what they were doing, how they were feeling, and so on. I was able to look at this stuff and I was able to show and demonstrate, like what kinds of days people experience when they would watch more at night and the answer to that was, they felt not very good at work.*

They had a really bad day at work. I found that people who didn't have a good tolerance for what I call unstructured time watched more TV. And I would say this is probably true for use of the internet too. If you don't feel comfortable with yourself and you need to fill up that time when you are by yourself and you don't have anything else to do, then what are you going to do? Well, you are going to gravitate to the easiest, most inexpensive medium that was ever invented, which is virtually free television. And fill up your mind with something else. So, it's kind of a form of escapism."

Little Light: *"The effects of television on the mind are probably one of the most studied aspects of our society and one of the things that takes place in the mind is that the frontal lobe is being bypassed. Within minutes* *of sitting down and watching television the frontal lobe activity simply goes almost to nothing."*

Dr. Neal Nedley, Owner of Nedley Clinic and Nedley Health Solutions: *"The frontal lobe of the brain is the seat of spirituality, morality and the will. It's actually the analytical portion of our brain and it's actually the decision maker. So, it's a crucial aspect in regard to our future success and happiness. It's how well our frontal lobe is functioning. And unfortunately, entertainment television suppresses the frontal lobe of the brain actually in about 90 seconds of viewing it, the frontal lobe circulation begins to go down and it actually has an adverse effect. You know the interesting thing is that people watch entertainment television often due to the fact that they feel a little depressed or anxious and it kind of calms them but in reality, it is a very short-term fix and it's going to complicate things in the long run.*

Of course, I'm most known for the one who treats depression and anxiety. In fact, we treat the most severe forms of depression and anxiety and what we have found is virtually every depressed patient will have about a 40 percent decrease in circulation and activity of the frontal lobe of the brain. So, we are trying to enhance the frontal lobe. Unfortunately, a lot of these gadgets that produce this over stimulation that can be fun for the short-term but you know there are more fun things to do than ever before in human history but yet we have more depression than ever before in human history.

More anxiety, more mental health problems are skyrocketing throughout the world and the entertainment medium is one of the primary reasons why this is occurring. Studies will clearly demonstrate that when you go to entertainment in a way to either get fun and it becomes a habitual process, the risk of depression and anxiety will more than double."

Little Light: *"So we decided to drive down to Los Angeles and talk to a top neurologist and find out really what's happening with the frontal lobe when we watch television."*

Dr. Mario Mendez, Behavioral Neurologist – UCLA Veterans Hospital: *"Well, each hemisphere has a prefrontal cortex, and it is, depending on how you measure it, 30 to 33 percent of the human neocortex. It has many functions which are often discussed in terms of cognition as executive abilities. I like to summarize it by having a role that you'd expect from a chief executive officer of a corporation, making strategic decisions, monitoring the performance of those activities, making corrections enroute and so on. It's also a source of social behavior. It turns out that humans are particularly visual animals. A lot of the information that is required for relating to the environment is visual and so a lot of the posterior part of the neocortex is devoted to complex visual processing beyond the primary visual cortex. I think of it as a gradual reconstruction. What you see is not necessarily a carbon copy of reality. What you see is a gradual coding and reconstruction of the signal in neurologic terms and in that process a lot of interpretation goes on."*

Little Light: *"Dr Mendez says the frontal lobe is like our command center. It's like where we make all these executed decision-making processes that our perception of reality is based upon, what we basically take in through our visual cortex. Now, question. Could you change someone's reality based on what you give them visually to watch? Dr. Michael Rich is a media expert from Boston's Children's Hospital wrote an article about what happens in your brain when you sit down, and you watch a 3D movie."*

Dr. Michael Rich: *"So what does your brain do when you're sitting in a theater, looking at a giant screen, wearing 3D glasses, swimming in surround sound, and processing the 24 images that flip by per second? Your brain dutifully processes those stimuli – and does little else. In fact, your pre-frontal cortex, which is involved in impulse control, future thinking, and moral choices, is basically inactivated in this process. That's part of why you 'get lost' in the movie."*

Little Light: *"Also researcher Jacob Jacobi found, after testing 5,400 viewings that 83 percent of those viewers misunderstood or* *miscomprehended that which they watched only moments before. That's what happens when the prefrontal cortex or the frontal lobe is shut off."*

Dr. Neal Nedley: *"You know, one of the interesting things about educational TV, you know for instance, C-span is educational TV, it may not be the greatest education but it's one camera view and there's a senator up there giving a talk and giving a lecture and you'll see something happen when you're in a room. You know, I, being a doctor, there's a doctor's lounge at the hospital I'm at, and often doctors look for educational TV when they're there together. And you will see them stand up and begin to vehemently argue against the television set. Because they'll say, you know what? He left this out, he left that out, he's not mentioning this, this is a very biased presentation. You will never see anyone argue with their television set when entertainment television is going on. You will see them actually just sitting there accepting all this, maybe laughing or crying, but they're not going to argue with the set because they don't have the ability. They may humanly disagree with what's going on, on that program, but they have no ability to actually bring forward that disagreement and start to logically go through the arguments in their mind because of the suppression of the frontal lobe of their brain."*[8]

In other words, you "won't" and "can't" even disagree with whatever "information" and "stimuli" and "programming" is coming your way! Again, you wonder why they call it TV "programming?" Again, that's exactly what these "Elites" are doing with it! They're "programming" your mind to go along with their nefarious agenda! To "believe" "act" "buy" "behave" "think" whatever they want! Television is simply the biggest global electronic "brainwashing" tool in the history of mankind, and we pay for it! No conspiracy theory here, it's demonstrated fact as you just saw! Again, as this person said:

"The question more and more concerning parents, psychologists and public officials is this: What is all this viewing doing to them?"

Gee, I'll tell you based on the facts… It's brainwashing us to "obey" "consume" "buy" "reproduce" whatever these "Aliens" I mean rich "Elites" want us to. So, now knowing what we just saw about Television, let's take a look again at that scene and see how every single aspect depicted in it really is going on. Check it out.

Clip from "They Live"

The guy walking down the street puts his glasses back on to see what the signs are telling the public without their knowing it. The big sign says, 'Obey'. He takes them off and on a couple of times to make sure he is seeing this correctly. He then looks at a sign advertising 'Come to the Caribbean' but when he puts on the glasses the sign is actually saying 'Marry and Reproduce.' He continues to walk down the street when he comes to a men's store. The sign on the store says 'Armani' but when he puts on the glasses it says, 'No Independent Thought' and the sign in the window says 'Consume,' not 'Close out sale.' When he puts the glasses back on, and looks down the busy street, he sees the sign telling people subconsciously to 'buy,' 'watch TV,' 'submit,' 'consume,' 'obey,' 'conform,' 'work 8 hours,' 'no thought,' and more.

He walks a little farther down the street and comes to the magazine stand and with the glasses on he can see that all these magazines also have the words on them as well. Even when he opens a magazine, he finds it's all printed out in there too. The place is saturated with all the words that can't be seen with the human eye without the special glasses he found. Then he has the experience of seeing the man in a gray suit come up to buy a magazine and when he puts on the glasses this man looks like an alien. He can't help but stare at this strange person. The alien asks him, 'What's your problem?' He asks him twice, but it doesn't seem that the human can speak. He is stunned that he can see this alien and doesn't say a word. The alien turns to pay for the magazine and then gets into his car and leaves.

The clerk comes over to him and asks if he wants to pay for the magazine he is holding. The clerk is holding money in this hand. With the glasses on the words 'This is your God' is printed on the money. He says, 'Look buddy, I don't want no hassle today, either pay for it or put it back.' He puts it back and continues to walk down the street. He passes a beauty salon and sees that there are aliens in there too. They are all over the place. And they are reading the magazines that have all those words on them in big letters. Not the regular articles.

His next stop is a little grocery store. As he walks in, he sees half the people in there are aliens. With the glasses off all these aliens look like regular humans carrying on normal conversations. He hears a voice coming from the TV saying, 'It's a new morning, fresh, vital, the old way is gone, we have faith in our leaders, we're optimistic of what becomes of it all, it all boils down to our ability to accept, we don't need pessimism.' This broadcaster is also an alien and the big letters behind him spell out 'OBEY.'[9]

Yeah, sure thing Mr. "Alien" I mean rich "Elite" who is using the whole Global Media system to "mesmerize the minds of the masses." I'll say it again. Looks like that science fiction movie has really become our everyday reality. How true it is and how clearer it gets the more forms of media we take a look at. Every one of them is being used against us on a massive global scale! But unfortunately, that's still not all. This "subliminal" "hypnotic" "addictive" form of Global Media called Television is not only being used to "control the narrative" of all information we're receiving around the whole planet to "mesmerize the minds of the masses" for these "Aliens" I mean rich "Elites" nefarious purposes, but it's also simultaneously creating the "immoral" and "scoffing" society that God warned about would also come upon the scene when you're living in the Last days! So, let's take a look at that unfortunate proof starting with the "immoral" aspect. Is Television really being used by these "Aliens" I mean rich "Elites" to encourage people to act ever increasingly "immoral" in the Last Days? You be the judge! Here's just a small sampling of what they're putting into our brains!

This clip opens with 4 family members sitting on a sofa all dressed in white. They are watching TV and eating popcorn. The movie they are watching is 'Gone with the Wind,' and they are watching the scene where Clark Gable says, 'Frankly, my dear, I don't give a damn!' But just as he says that word, the boy on the couch gets hit with a blob of red paint. What is happening is that every time a bad word is expressed on TV, they are hit with a blob of paint from a paint gun.

'Gone with the Wind' had the first swear word in cinema history in 1939. In 2013 'The Wolf of Wall Street' had the most swear words in cinema history.

F-words: 528
Blasphemy: 70
Other language: 200
Sex/Nudity: 27
Graphic violence: 3

The family on the couch get black bags put over their heads because it might get a little dangerous when the movie begins. The guns are all lined up to begin shooting. For every bad word or bad thing that happens, they are shot with a blob of paint.

Hundreds of shots are hitting them. Dishes are breaking, popcorn is flying, pictures are being blasted and it keeps coming. They are getting covered with all the paint. Finally, it is over, but every word has impact. 'Protect yourself and your family' is the message.

Next clip:

Deep in the center of the American suburbs was a 15-year-old girl named Jenny. By all appearances she is an average, healthy high school student but at the heart of her tragic story we find circumstances that are widespread across her peer group around the country. Jenny awakens every morning in her room, alone, with only a twin sized bed. Her request to her parents of a larger, newer, bed continues to fall on deaf ears. Even her cry for bed linens with a higher thread count, continue to go unanswered.

In rural Southern Africa, Annie lives with her grandmother and her two little sisters in a one room hut, that has been badly damaged by storms. She sleeps on a grass mat on a hard dirt floor. There is only one blanket for everyone to share.

Sadly, teenagers like Jenny, only have an average of $267.00 to spend per month. That is a little more than $65.00 for an entire week. At breakfast Jenny usually has to fight with her siblings to get the last pop-tart in the pantry. On the days she loses the battle she is forced to eat cereal and fresh fruit. The drive to school every day creates another clash with the siblings for the coveted location in the car known as shotgun in the front passenger seat. Even when Jenny is victorious, she finds herself cold, as the leather seats of her family's BMW lacks seat warmers. To make matters worse, Jenny's mother makes frequent attempts at carrying on an actual conversation with her daughter unleashing a barrage of small talk upon Jenny from the driver's seat.

Annie and her brothers work in the fields to help their grandmother grow corn for their meals, however their country is suffering from a poor year's drought and this year's crop will not yield enough food to carry them through the year.

In Jenny's closet she suffers a serious lack of brand names sewn into her dozens of shirts, jeans and dresses. Shockingly she was even made to choose just one purse from Fendi, Gucci, and Louis Vitton at the mall.

Annie and her sisters have two dresses each and none of their clothes fit them well. They have holes and are thread barren in some places. After a long day of working in the field, the girls spend their time helping their grandmother repair their dilapidated hut or digging holes in the dry riverbed in search of water.

Finding herself in another no-win situation, Jenny does her best to remain brave while choosing between tap water and wholesale bottled water to wet her parched mouth.

When Annie and her sisters do find water in the riverbed, it's usually dirty from cows, donkeys and other animals drinking from the same water hole. There are also countless bacteria and other organisms in the water that could make them very sick.

Don't become a victim of teenage affluenza. You only have one life, do something.

Next clip:

And the spirit lifted me up between the earth and the sky and brought me the vision of God to Jerusalem to the North Gate of the Temple. Then the Lord brought me to the entry of the Courtyard and when I looked, I saw holes in the wall. 'Now dig into the wall,', he said. I did and uncovered a door to a hidden room. Then the Lord said to me, 'Go on in and see the terrible and evil things they are doing.' So, I entered and looked, and the walls were covered with pictures of all kinds of demonic and evil things.

All the various idols that were worshiped by the people of the land. There they were standing before the images fixing their minds on them and lifting up their hearts to the things, they worshiped creating a thick dark cloud above their heads. Then the Lord said to me. 'Son of Man do you see what the people of the land are doing in the dark. Each man with his idols?' And they say, 'The Lord does not see us, He has gone away.' **Ezekiel 8:3,7-12.**[10]

Oh, He has not gone away, and He does see what we are putting into our brains. And He also warned about the side effects. Junk in equals junk out! You ingest hours and hours on end of "immoral" content on TV, you're going to act like it. You're going to become an "immoral" society. How many times have you heard people say, "Why are people acting so immoral these days? Why is it on the rise? Why is it getting worse and worse?" Well, you just saw one of the biggest reasons why! It's Television! And if you need even more proof, add to what you just saw, these sobering statistics.

- The average American adolescent will view nearly 14,000 sexual references on TV per year.
- 75% of prime-time network shows included sexual content, up 67% in one year alone. Nearly one-third of family hour shows contain sexual references.
- Now they're offering full-blown nudity shows. "Seven shows right now are being rolled out that feature complete nudity." Not just showing nudity here and there, that's bad enough, it's complete full-blown nudity all the time.
- MTV has a couple of new shows coming out, one is called "Virgin Territory" where participants are trying to lose their virginity or what they call V-Card. Another show is called, "Happyland" where there a teen story line that promotes incest. And the lead person playing the girl in the show said, "Incest is hot and we're going to have fun!" on TV!
- Now they're airing commercials on TV promoting adultery!

CNN Prime News: *"Marriage is hard enough, and we certainly don't need TV commercials to try to get us to cheat on our spouse. But they are out there. In your face ads that even your kids can see them. Take a look!"*

Affair Moments by Ashley Madison: There is a couple walking down the sidewalk, over his head the word, HIM and over her head ON A BUSINESS TRIP. The next couple that is shown has over her head, HER and over his head, WITH HIS BEST FRIEND. Another couple that looks like they could be in a church, over his head, HIM and over hers, WITH A WORK COLLEAGUE. Several other couples are shown cheating with the opposite friends, co-workers, neighbors, being advertised online with almost nine million members. Isn't it time for AshleyMadison.com? Life is short. Have an affair.

Music is playing and someone comes on singing, 'I'm looking for something new, something for myself, something that I've got to do. I can't wait to decide! There's nothing that I haven't tried. Shhh... Ashley Madison.

During the song a man pulls up in his car and picks up a girl on the side of the road and they drive to a hotel.

Life is short, have an affair!

CNN Prime News: *"Life is short, have an affair, that's the tagline for AshleyMadison.com. An online dating service for married people, advocating adultery."*[11]

- The average American child or teenager views 10,000 murders, rapes, and aggravated assaults per year on television.
- 80.3 percent of all television programs contain acts of violence.
- A child born today will witness 200,000 acts of violence on television by the time they are eighteen.

I'll repeat what the one researcher said:

"The question more and more concerning parents, psychologists and public officials is this: What is all this viewing doing to them?"

I'll tell you. It's causing the massive increase in "immoral" behavior that we're seeing today that God warned about would come on the scene when you're living in the Last Days! But that's only half of it! These same "Aliens" I mean rich "Elites" are also using the Media of Television to create the biggest "scoffing" society in the history of mankind against God, Jesus, the Bible, and Christianity! See if you agree. Watch this "scoffing" evidence.

Narrator: *"The two Christian film offices that had been active in providing moral guidelines for the entertainment industry for over 30 years were shut down as well. New ideas and a new breed of people began to take their place. Among them was Anton Szandor LaVey."*

Anton LaVey: *"We believe in greed, we believe in selfishness, and all of the lustful thoughts that motivate man because this is...."*

Narrator: *"After opening up his church of Satan in San Francisco, LaVey began to advise filmmakers involving matters of the occult. LaVey's satanic beliefs make his observations about the film industries new attitude towards Jesus very interesting."*

"The satanic age began about the year 1966 and now we are beginning to see in 'Jesus Christ, Superstar' and 'Godspell' Jesus as fallible, no different from any other man – which was originally what Satan was supposed to represent." **Sherry and Brad Steiger, Hollywood and the Supernatural, St. Martin's Press, 1990, p.13."**

"LaVey proved to be far more discerning than the multitude of church going people brought into the dumbing down of the Messiah."

Clip from "Fletch", Universal Pictures: ##### ###### on a Popsicle stick.

"The Favor, The Watch and the Very Big Fish," Films Ariane/Umbrella Films, LTD, show Jesus hanging on an upside down cross.

Sam Kinison: "Banned" – Giant Records

The Dream Team, Universal Pictures: "This is the body and blood of our Lord Jesus Christ" being said by a man sitting in a chair holding up a liquor bottle.

The Crow, Dimension Films/Miramax: "Jesus Christ walks into a hotel, gave the innkeeper three nails, can you put me up for the night."

When asked about her venomous hatred of Christ in the Ken Russell film "Laire of the White Worm, Donahoe replied, "I am an atheist, so it was actually a joy. Spitting on Christ was a great deal of fun." **Interview, Sept. 1991, p. 109**

From cartoons like Beavis and Butt-Head, MTV/Viacom to comics, Robin Williams, Paramount, "The devil is God when he is drunk."

Kids in the Hall, CBS, "God did exist, He died, He was very small, mystery solved."

Mosquito Coast, Warner Brothers: Whether it is showing contempt for scriptures.... The Bible is tossed in the air, and the person throwing it says, "Just what I have been warning you about!"

Roseanne, ABC: Poking fun at going to church.
Daughter: "It's even worse than you thought."
Dad: "Well, what could be worse, he's not doing drugs is he?"
Daughter: "No, No, he's going to church."
Dad: "Oh God, No!"

Picket Fences, CBS:
Child: "I'm off to church."
Brother: "He's wacked."

The Simpsons, Fox Broadcasting Co.: "You want to say grace?"
Son: "Dear God, we pay for all this stuff ourselves, so thanks for nothing."

Dances with Wolves, Orion Pictures: With the pagan practices of the American Indians, they are treated with loving respect.

Little Buddha, Miramax Films: Eastern religions bask in the warm fuzzies of spiritual awe.

Out on a Limb, ABC Video Enterprises: Even the occult is given the benefit of the doubt.
Shirley MacLaine: "He said the divine force is what the soul is made of."

Handmaid's Tale, Cinecom Entertainment Group: While the Christian faith of untold millions is portrayed as a potential breeding ground for Neo-Nazis, public executions and a religion induced madness that can lead to everything from mob violence to a mother's willingness to kill her child.

The Rapture, New Line Cinema:
Mother: "Do you love God?
Daughter: "Yes."
Mother: "Pray baby."

All too often this is the gospel of Hollywood.

The third of the Ten Commandments, given to us by God, warns that we should treat even the mention of His name with great awe and respect. Never using it in a casual or disrespectful way. The Old Testament saints took this command very seriously, insisting that scribes wash their hands and use a new pen when even writing the name of God. Many of the Jews were even afraid to say it out loud in any circumstance. Choosing to use it indirectly as 'the name.' Contrast that to our day, to a time that God has an incomprehensible grace and mercy has chosen to reveal himself in Christ as Immanuel, God with us. The true name of the Lord our God.

"The only name under Heaven given to men by which we may be saved."
Acts 4:12

God's name is degraded, blasphemed, and reduced to a cheap expletive device by so many in the entertainment industry that wonders if it isn't a default key on their word processor.

"Sometimes a Great Notion" by Ken Kesey and the academy award winning "One Flew Over the Cuckoo's Nest, Fantasy Films/HBO provided perhaps the best summation of the pride and irreverence that characterizes so much of today's art and entertainment. 'The job of the artist is to say, 'F### you, God! F### you and the Old Testament you rode in on!'" **Esquire, Sept. 1992, p. 210**[12]

And you still don't think that Television is being used to encourage people to "scoff" at the things of God? Even the industry leaders admit it, when they're being honest, as brutal as it is! Again, this is no conspiracy. It's really what's going on and going inside your brain every time you turn on that "manipulative device" called TV. It reminds me of the following acronym one person emailed me:

MEDIA
M –Most
E – Effective
D – Devil
I – In
A – America

Not to mention, around the world! In fact, by now, it probably feels like you're getting beat up, literally, with all these shocking revelations, one after another, of how everything you "read," "hear" and "see" today really is a massive global Subliminal Seduction by "Aliens" I mean rich "Elites" to "control" and "manipulate" you. And I'm sorry we've had to "punch you in the gut" several times with all this information, but it really is for your own good, like this scene portrays.

As the main character (MC) walks down the street he looks down an alley and sees a bald man in a blue sweatshirt standing in front of a dumpster. The man turns to go in the opposite direction and the (MC) calls out to him to wait. As the man turns around, he yells out, "You better find yourself somewhere to hide and keep praying that nobody ever finds you!"

He holds out the glasses that he found and tells the guy at the dumpster, "Here try these on."

In response the bald guy says, "Stay away from me!" But the (MC) walks towards him holding out the glasses for the bald guy to see what he had found and what they were showing him. He says, "I'm telling you, you don't ...." Before he could finish his sentence, the bald guy punches him in the nose. Now he is bleeding. He composes himself and tries to warn this guy, "I am trying to save you and your family's life."

The bald guy responds, "You couldn't even save your own!" Then the (MC) walks up to him and punches him in the nose. He says, "I'm giving you a choice, either put on these glasses or start eating that trash can." The bald guy responds, "Not this year!" The (MC) says, "Okay!"

As they start to fight in the alley, it seems the bald guy knows what he is doing. He gets in 3 or 4 punches before the (MC) can get in one and he finds himself laid out on the ground. He stands back up and as he has new energy to fight back. The bald guy is on the ground this time and is starting to get up when the (MC) is telling him, "I don't want to fight you!"

But that doesn't stop the guy from charging at him again. They are running out of steam, when the (MC) finally knocks the bald guy to the ground. He is having trouble getting up when the (MC) gives him a hand, pulls him to his feet, and then hands him the glasses and he says, "Put them on!" At that point, the bald guy throws the glasses to the ground, raises his foot and is about to smash them to pieces when the (MC) yells, "No!" He pushes the bald guy back, so the glasses wouldn't get smashed but again the bald guy knocks him down again. He picks up the glasses and throws them against the wall and starts to walk off again.

The (MC) gets back on his feet and starts to run after the bald guy and pulls him to the ground one more time. The fighting doesn't stop. They are about to kill each other. When they get to a point where they are both laid out in the alley, the (MC) manages to get up, pick up the glasses, and put them on the bald guy, making him stand up and look out at the street. The bald guy sees these aliens are all over the place. The (MC) says, "Look, look at them, they are everywhere! Now hold on. You're not the first one to wake up out of their dream."[43]

And now you are, because that's what we're trying to do for you with this study. And I know it's tough, trying to take in all this information exposing how these "Aliens" I mean rich "Elites" really are "mesmerizing the minds of the masses" on a global scale using all these various forms of Media we've been looking at. But don't "take the glasses off" now. And certainly don't "pass out" from all these punches we've been throwing your way. We are almost there, hang in there, keep them on because we've only got one more to go.

But as you can "see" it's the same pattern with all these different forms of media. They are all being used to "control the narrative" and to "manipulate the minds" of the masses for the "Elites" "own" nefarious purposes and it's going on all over the world, whether you realize it or not! Including, once again, the "desire" to "manipulate" the "attitudes" of people, specifically against God and Christianity, so they will "scoff" at Him on a global basis and create an "increase of wickedness" in people's behavior like the Bible warned about in the Last Days. Total "manipulation" of the "minds of the masses," across the whole planet, 24 hours a day, 7 days a week, just by simply "owning" the media called Newspapers, Radio, Music, Books, the Education System and now even Television. Right now, as we speak, the whole planet is being bombarded by all six and there's no escaping it!

But unfortunately, it's still going to get worse. These "Aliens" I mean rich "Elites" are not merely content with just owning all the Newspapers and Radio Broadcasts and Music Industry and Books and Education System and Television outlets around the whole world to "control the minds of the masses." In their insatiable lust for "power" and "domination" they are literally out there amassing "all forms of media" in order to get the job done. Including the latest one they've invented to further "manipulate" us. Of course, I'm talking about Social Media. Believe it or not, that too is being used by the "Aliens" I mean rich "Elites" to "indoctrinate" the minds of people into what they want them to "think," "act," "believe," or "do," which will be the topic of our next section. And as we will soon see, Social Media is also telling us "to not think for ourselves," "to obey only what they want us to obey," "to never question them," and of course, "to scoff" specifically at the Bible, God, Jesus, and His Soon Coming Judgment! Where have I heard that before? I know it's hard to believe, but once again, hang in there for one more round of punches, "keep those glasses on," and let's begin this final journey and expose how this next form of "media" is also "mesmerizing the minds of the masses."

Chapter Seven

The Manipulation of Social Media

The **5th way** the global media is "controlling the narrative" to "control our minds" is with **Social Media**. You see, Newspapers, Radio, Music, Books, the Education System, and Television are not the only Subliminal Manipulative Technologies being used on us on a global basis by a small group of "Aliens" I mean rich "Elites" around the world. Believe it or not, the recent trend of Social Media is as well. It too is out there "swaying our minds" subconsciously "telling" us to "buy" and "consume" and even "instruct" our minds "to go back to sleep," "never question authority," "obey" and "believe" whatever these rich "Elitists" "want us to believe" including the need to "scoff" at the things of God.

So, let's "keep those glasses on" one more time and "see" how this next Media Technology called Social Media is being used to "manipulate the minds of the masses" on a huge global scale. And not so surprisingly, just like all the other forms of Media we've been examining, this "manipulation" that's being done, as we speak, with Social Media, has been going on, pretty much since its inception, whether people realize it or not. Let's "see" how it all began.

"It's hard to imagine a world without social media. In today's society we use social media for a variety of purposes, to make new friends, keep up with celebrities and brands, to promote ourselves, express our thoughts and opinions to the world or to stay in contact with friends and family who we do not often see, especially those who live a long way away.

Social media has only been around for about two decades. Before that, methods of staying in contact with friends and loved ones over long distances away were not merely as simple. Let's take a look back at some of the earlier methods of communication starting all the way back in 550 BC. Written correspondence delivered by hand from one person to another was the earliest method of communicating over long distances.

The earliest postal system was created in 550 BC in ancient Persia. The Persian King Cyrus the Great mandated that every province in his kingdom would organize the delivery of posts to all of its citizens. Several other postal systems were introduced in many other countries following the Persian system; however, Rome established the first well-documented postal service. Organized at the time of Augustus Caesar, mail was delivered by light carriages pulled by horses. This service was created for government correspondence but another service for citizens was later added.

Over many centuries postal delivery systems became much more widespread and organized. These systems are still widely used today but thanks to technological advancements there are other faster ways to communicate with loved ones over long distances.

The telegraph was developed in the 1830s and 40s and the first telegram was sent in 1844 by Samuel Morse who is also known for his invention of Morse Code. The telegraph allowed messages to be delivered over long distances much faster than through a postal system. Telegraphic messages had to be kept very short, but they were the fastest means of long-distance communication at the time and were one of the most significant advancements in communication technology that had occurred in centuries.

In 1876 Alexander Graham Bell invented the telephone. Telephones were a revolutionary technology. They allowed people to communicate across long distances instantaneously which had never been the case before.

In the 20th century technology began to advance at a rapid pace. The first supercomputers were created in the 1940s and then as technology developed, engineers began to create networks between computers. Personal computers made their first appearance in the 1970s and continued to develop further with advancing technologies. It was J.C.R. Licklider who wrote one of the first descriptions of what would later become the internet."

"A Network of such (computers), connected to one another by wide-band communication lines which provided the functions of present-day libraries together with anticipated advances in information storage and retrieval and other symbiotic functions." J.C.R. Licklider, Man-Computer Symbiosis, 1960

"Licklider began development for one of the computer networks in the internal network for the United States Department of Defense called ARPANET, the first network link established in 1969. Several other networks followed including Usenet, which allowed users to communicate through a virtual newsletter.

In 1978 the first email system was developed which allowed people to send and receive messages of text instantly. The main difference between the telephone and email in terms of long-distance communication is that with the telephone both parties had to be available at the same time whereas email messages did not need both sender and receiver to be available in order to work. They could each read and send messages when they had access or when it was convenient for them.

The system that would become the internet emerged in the late 1970s and its initial purpose was to connect all the various networks that had been created and unify them. In the late 1980s the first internet service provider *companies were created, and the internet was first opened up to non-government use."*

"Every day, America Online is making it easier for people to live, work and play."

As his friend is sitting at the computer his friend asks him. "Hey, Dan, ready for the game? I'm just finishing up here with my new kayaking friends."

Dan asks his friend, "Kayaking friends on your computer?"

His friend answers: "Yeah, I just got America Online."

"The Worldwide Web was invented in 1989 by Tim Berners-Lee who also created the first web browser called Worldwide Web. Until Social Media came along, the internet was mainly used for mailing lists, emails, ecommerce, online forums and personal websites.

Techquickie: *"People were already sharing content online through Flora, Chatroom, and Tripod pages. But what was the first site that used the modern paradigm of creating profiles and sharing different kinds of information with a subset of sites of users that you were connected with somehow? A number of sites could try to claim that title, but a strong candidate was SixDegrees, which entered the scene in 1997. It featured plenty of elements that are now staples of social media, such as friends list and instant messaging. But unfortunately, they closed their digital doors in 2001 because this is basically the quintessential example of something being too far ahead of its time, there just weren't enough people on the internet back then to keep the site afloat.*

It wouldn't be until five years after SixDegrees was born that the first social networking service hit a million users. Friendster, launched in 2002 quickly became widely popular, by showing users, not only how they were connected to others but with a slicker presentation, similar to online dating sites of the time. And they also benefited from the shrinking stigma around meeting people online due to the Internet's explosive growth. But while Friendster quickly boasted a user base of over three million people, tough competition was on the horizon.

If you are around the same age as I am (31 years old) you probably fondly remember or at least remember, MySpace. MySpace took a great deal of inspiration from Friendster. But because the folks behind MySpace already ran a company called eUniverse which operated multiple websites, like gaming and dating, MySpace had the advantage of being able to get the word out through eUniverse's existing user base. This helped them to rapidly become more popular than Friendster.

In fact, Friendster's founder claimed that MySpace employees made posts on Friendster forums asking visitors to come join MySpace instead. MySpace was also able to develop new features for their site faster than Friendster. Meaning, for a while, that the internet was awash in profile pages featuring strange backgrounds and weird sparkling gifs. And real life was awash in novelty T-shirts directing onlookers to said pages (Yes, I'm on MySpace).

Of course, MySpace's dominance was also short lived thanks to a famous Harvard dropout. Mark Zuckerberg developed Facebook as a sophomore in 2004. He originally intended it to be a campus wide student directory. Apparently, because he was simply unhappy that it was 2004 and the University still hadn't created one. After over a thousand Harvard students registered to the site the first day, Facebook started expanding to other schools, requiring a University or College email address in order to sign up.

By the fall of 2005 most Universities in the US had access to the site. Its explosive growth fueled by the ease with which the students could see who they were connected to through classmates that they already knew. This early version of Facebook lacked modern features, acting as a digital replacement for university directories with much more emphasis on showing background information through user profiles. And the only way that visitors had of submitting anything to another user's page was by changing the wall. A block of text that any user's friends could edit at will.

Early on, Facebook didn't have the same pull as MySpace with advertisers but changed once Facebook opened itself up to the masses in 2006. For better or for worse Facebook quickly dwarfed MySpace. The most prominent reason for this being the two sites diverged in terms of their functionality. Facebook's model was focused much more heavily on the user generated content and ways to make the networking part of social networking easier, pages, groups and content sharing.

MySpace meanwhile was widely considered to have spread itself too slim with extraneous features like karaoke that didn't work very well and a design that was quickly becoming clunky, slow, and outdated. MySpace also served their users with tons of ads and became susceptible to spam and malware. Facebook was regarded a much better maintained with a cleaner design and a very obvious influence on other social networks like VK in Russia and RenRen.com in China.

So, in 2009, MySpace laid off 600 workers and after a couple of sales, it is still around but as a music centered platform that is no longer trying to compete as a big time do it all social media service. Meanwhile, while Facebook and MySpace were grappling for social networking dominance Twitter tripped onto the scene in late 2006. And although the social media space was already becoming crowded, Twitter stood out because it wasn't a copycat site.

Instead of focusing on user profiles and a huge set of social networking features, it was based on a much simpler idea. One hundred and forty characters, micro-logging posts or tweets, that could be used to reach everyone on the site instantly. This simplicity made it popular with convention goers at the 2007 edition of South by Southwest leading to some serious publicity and an explosion in usage going from twenty-thousand tweets per day, before the convention, to fifty million in 2010, and of course the introduction of the often-misused hashtag in 2007 which Twitter turned into hyperlinks in 2009 and started to use to determine what topics are trending in 2010. It was this addition of trending to Twitter and later on Facebook that represented the next big trend on the major social networking sites.

With many turning at this point to trending topics and their customizable feeds instead of traditional media to get their breaking news. This drew in the extra advertising dollars that have allowed services like Facebook and Twitter to expand into areas like live broadcasting and embedding richer forms of media directly into the posts. The fact that the way that we interact with each other on these social media sites has become mainstream means that we actually have even seen these social features of one kind or another implemented on other sites. Like how LinkedIn, a business networking site which has actually been around since 2002, now looks decidedly Facebook like. Trending and community functions have shown up on YouTube, hashtags are on Instagram, now a Facebook subsidiary, and Reddit has a vote-base system which is roughly an allegorist to likes on other platforms."

"Today there are a huge variety of social media sites for various purposes, however the main purpose of social media continues to be connection to friends, family and strangers.

For many, social media is a way to stay in touch with lifelong friends and keep up with family who live in other states or countries and to share photos, videos, news, thoughts and opinions or even just to send a quick message to say hello."[1]

As well as allow these "Aliens" I mean rich "Elites" to "manipulate the minds of the masses" on a scale never before dreamed of! And if you don't think that's what they're doing with Social Media around the planet, let's look at just one of them, that of Facebook. From its inception, Facebook was designed not merely for the users to "share" information, but rather to "acquire" information on the user themselves. Don't believe me? Let's see just where Mark Zuckerberg got his "seed money" to launch Facebook in the first place.

"Do you have a Facebook? Have you thought about the privacy you put at risk? Facebook allows users to post their favorite music, books, movies, their address, hometown, phone number, email, clubs, jobs, educational history, birth dates, sexual orientation, interests, daily schedules, exactly how they are related to friends, upload pictures of themselves and even political affiliations. It's privacy policy even goes so far as to state it 'also collects information about you from other sources, such as newspapers and instant messaging services. This information is gathered regardless of your use of the web site.'

Have you seen the Facebook's pulse feature? Pulse provides statistical trends among universities down to minute details, such as percentages of females with conservative views, the student body's top ten movies, and the percentage of students who have read 'Catcher in the Rye.'

The so-called privacy policy goes on to say that they share your information with third parties including responsible companies of which they have a relationship. Can you think of any marketing group who could pass up such valid yet easily collected statistics such as these and others? So maybe they're using us. But is there more?

Funding came in the form of $12.7 million dollars from Venture Capital Firm, Accel partners. Sales Manager, James Breyer, was former chair of the National Venture Capital Association. Breyer served as Venture Capital Association Board with Gilman Louie, CEO of In-q-tel, a Venture Capital Firm established by the Central Intelligence Agency in 1999. This firm works in various aspects of information technology and intelligence, including most notably nurturing data mining technologies.

Breyer has also served on the board of BBN technologies, a research and development firm known for spearheading the ARPANET, what we know today as the internet. The IAO (Information Awareness Office) stated its mission was to gather as much information as possible about everyone in a centralized location for easy perusal by the United States government including but not limited to the internet activity, credit card purchase history, airline ticket purchases, car rentals, medical records, educational transcripts, driver's licenses, utility bills, tax returns and any other available data. All of the above raises more questions than answers."²

      Yeah, I would say so! Boy, have we been duped! You just thought Social Media was for us to "connect" with other people. Silly rabbit, it's all about building huge, massive, mega databases on us to create a detailed psychological "profile" on each of us, on a global scale, not to mention also track our "faces" wherever we go! I mean, after all, they call it "Face" book. Yes, your "face" is "booked" alright! It's in their databases tracking and building a massive global "profile" on you concerning your every move. Your likes, dislikes, contacts, interests, patterns of behavior, beliefs good, bad and ugly, you name it! It's all being stored, tracked, and monitored! And they're not the only ones! All major Social Media entities are also out there using their platforms to build huge, massive, mega global databases on us. And for those of you who need more proof of that, here you go. Again, let's just for now focus on the single Social Media entity called Facebook. But, keep in mind, they all do it.

Kiran Chetry, CNN Reports: *"Well if you or your friends post pictures of yourself drunk or other compromising positions on Facebook, you will want to hear this. A change to the company's fine print is raising questions online this morning about what happens to all the information that you post on your page. Alino Cho is following this story for us. First of all, you shouldn't post pictures of yourself in compromising positions anyway. But there is something interesting that has people scratching their heads. Apparently, a blogger found a checklist."*

Alino Cho: *"Essentially we are talking about the TOS, Terms of Service, in respect to Facebook. It is probably something that most people who use Facebook have not heard about. About two weeks ago, the fine print explaining privacy rights on Facebook quietly and subtly, changed. Here's what happened. Facebook removed language that said its ownership of your content would end when you remove that content or close your account. Now what does that mean? It means that Facebook now would continue to have access to things like your personal photos and personal information. The blogs, as you might imagine, went crazy. The*

Consumerist, as published by Consumer Reports, responded by saying Facebook's new terms were tantamount to 'We can do anything we want with your content, forever.'"

"Facebook can take all of the data that you submit and combine it with data from other users and outside information to construct a profile of you. Facebook uses nearly a hundred different data points to classify your interests and activities. This would include basic stuff like your age and gender, but also more complicated information, like whether you own a motorcycle or you recently went on vacation or whether you are a gadget geek. Researchers have found that by using signals such as your likes or actions, Facebook can tell if you are in a relationship or going through a breakup.

Facebook doesn't just know who you are, it also knows where you are, if you have location tracking turned on. Facebook collects an enormous amount of location data about where you are going, where you came from or where you live, where you work, restaurants and businesses you tend to go to and they use this information to target ads at you and location data can reveal other people who live in your house, even if you're not connected to them on Facebook.

Now obviously Facebook knows what its users buy when they click on ads on Facebook. But what most people don't realize is that they have ways of tracking your offline purchases as well. For many years Facebook has had partnerships with data brokers that collect information about people's purchases. So, for example if you buy a burrito with your credit card Facebook can know about that transaction, match it to the credit card that you added to Facebook or Facebook Messenger and start showing you ads for indigestion medicine.

One of the most controversial parts of Facebook data collection is a feature called People You May Know. This is where Facebook uses many different signals of what it knows about you to determine who else you might be connected to. And this is not always things you share with Facebook, it might be contacts in your phone, it might be people who have been in the same room as you, Facebook has been using location data to recommend friends. So, it might have been recommending people that shared a doctor with you, or work in the same building.

Facebook can also be used to compile data about your political activity, like protests or marches you go to. In one case in 2016, the ACLU found that 500 police organizations had signed up for a service called GOPDF which scraped data from social networks like Facebook, Twitter and Instagram to help officers look for users who might be in a specific location or attending a specific protest.

Facebook doesn't just know who you are or where you are and what you buy, it also can be used to figure out what kinds of things you might do in the future to predict life outcomes, like whether you will be addicted to substances, whether you will switch political parties, whether you are physically healthy or physically unhealthy."

Juju Chang, ABC News: *"Science Fiction almost got it right. Face recognition technology has become part of daily life. But instead of controlling security access to top secret facilities the technology is being used to identify the people in that crazy picture from last weekend's beach party. It's all happening, of course, on Facebook, which has been quietly using face recognition software on every photo you upload."*

An ABC Reporter/Facebook user: *"Facebook knows me, it knows my friends, it tracks what I like, it knows where I live, it knows my sisters, Christy and Laura, but now Facebook also knows my face. My face out of everyone else's. Not just mine, but if you're on Facebook, Facebook can recognize your face too. Here's how. While we have been uploading nearly 20 billion photos, Facebook has been making file metric fingerprints of all our mugs. Now Facebook has a massive collection of our faces. Probably the largest collection in the world. Kind of cool, kind of creepy. Here's how it works.*

Intern Brenna Williams: *"I'm going to upload some photos of myself..."*

ABC Reporter/Facebook user: *"She is one of millions of people who are using this feature already. When she uploads photos, Facebook recognizes her and her friends and says, 'Hey, that's you, right? Most of the time it is, and tagging becomes almost automated. Face recognition technology has been around for a while. Mostly used by police. They even scan people walking into the Super Bowl and caught 19 criminals. The guys at Face.com in Israel, showed us how Facebook's technology works. They are the largest facial recognition company online. Once they had my photo, the computer recognized me, no matter where I was every time.*

This got us at Nightline thinking, could we trick Facebook? I tried one other way to confuse the computer. (He put on a ball cap, large sunglasses, and a piece of tape on his upper lip.) We sent that photo back to the guys at Face.com and they recognized me even in my cover-up. Which goes to show that with that disguise you may fool your friends, but it probably won't fool Facebook."

FOX29 Reports: *"Did you hear about this? Is Facebook playing another Big Brother. Some users are saying that the social media site is really crossing the line with their new Messenger app. Joining us now is our new Tech expert Anthony Mongeluzo. Anthony, thanks for joining us. So many people are talking about this Messenger app that used to be optional but now it's going to be mandatory, I guess. You have to download it. How does it all work and why are people worried about it?*

Anthony Mongeluzo, Tech Expert: *"Well, basically, this new app is getting terrible reviews. It has a one out of five stars on the iPhone app store and similar ratings on the Android. What this Messenger app basically does is you have to give away all of your privacy. You allow Facebook to look at your call list, at your contacts, your text messages, it can even use your recording device and your camera device on your phone without even telling you."*

Fox29 Reports: *"Okay, that's the thing that scared me the most. Take me through that. I'm holding my phone right now. If I have this Messenger app, they see what I see or even worse, spy on me with what I'm doing?"*

Anthony Mongeluzo: *"Most of your phone now days have two cameras, one in the front and one in the back, so legally you are giving them access to those cameras when you install that application."*

Fox29 Reports: *"So some creepy guy in the San Francisco Bay area, someplace in Facebook, can be looking live at what I am looking at?"*

Anthony Mongeluzo: *"Don't just worry about San Francisco, Facebook is global, around the world, they can look at you."*

RT Reports: *"What has Facebook been up to, why does it even affect people not signed onto Facebook? Web developer and surveillance expert, Sander Venema, Founder, Asteroid Interactive, joins me via Skype from the Netherlands. Sander welcome back since going underground. So, what has Facebook done to trample over European privacy laws?"*

Sander Venema: *"There was a report published by the Belgian Privacy Commission about Facebook's tracking behavior and more specifically the social pluckings that they use. These are the 'Like' button and 'Share' buttons that see millions of websites across the internet, and these are loaded in from Facebook's domain name that basically places a cookie, Facebook places a cookie on your computer, whenever those buttons are loaded in."*

RT Reports: *"So if you press 'Like' or one of the 'Like buttons, which doesn't even have to be on your Facebook site, it can be anywhere, it could be asking your ideology or your political persuasion, it all gets stored on your computer and this is what the European regulators are worried about?"*

Sander Venema: *"Actually before you click 'Like' and when the 'Like' button is loaded in, so you load up the website, it has the 'Like' button on it before you even click on it, information went to Facebook and then Facebook places the uniquely identified cookie on your computer which they subsequently use on all of their other sites with Facebook like buttons to map the sites that you visit basically and map your interests."*[3]

And everything you could possibly be doing while you're online or on your phone! This is precisely why I call Facebook for what it is. It's not only "face" book but it's "trace" book because that's really what's going on here folks, as you just saw! On the one hand, they're not hiding it. I mean, think about it. Social Media accounts are called "profiles," right? And that's exactly what they're doing. They're building huge, mega, massive global "profiles" on every aspect of our everyday lives whether we realize it or not, of everything you can think of, what we are doing at any given moment! Including "cataloging" all of our friends, likes, dislikes, contacts, interests, patterns of behavior, beliefs, good, bad and ugly, you name it! It's all being "cataloged" and "profiled."

So that leads me to the next question, "Why? Why would Social Media entities be doing this to us? I mean I thought it was just for having a fun and creative way to "connect" with people across the whole planet. I had no idea they were going to use it to "connect" "me" to their "databases" and "profile" every aspect of my life across the whole planet!" Well, believe it or not, it's not only being done for the obvious "monetary" reasons, which are self-evident by the way. These "Aliens" I mean rich "Elites" are raking in billions, not millions, but billions of dollars every single year off of all this information they're gathering from us. In fact, here's the Top 10 richest Social Media billionaires on the planet currently.

- 10) Hae Jin Lee, of Line – $1.7 billion.
- 9) Reid Hoffman, of LinkedIn – $1.9 billion.
- 8) Sean Parker, of Napster – $2.7 billion.
- 7) Evan Spiegel, of Snapchat – $4.9 billion.
- 6) Jack Dorsey, of Twitter – $8 billion.
- 5) Jan Koum, of WhatsApp – $9.9 billion.
- 4) Zhang Yiming, of ByteDance – $22.6 billion.
- 3) Ma Hueateng, of Tencent – $55.3 billion.
- 2) Larry Page, of Google – $64.4 billion.
- 1) Mark Zuckerberg, of Facebook – $92.9 billion.

So, as you can see, there are huge "monetary" reasons why these Social Media giants are gathering all this giant amount of information on us. They are raking in billions and billions of dollars every year off of it, through advertising, targeted advertising, buying, selling, subscriptions, services, fees, you name it! However, these Global Social Media giants are not only using their Media platforms for "monetary" reasons but also for "monitoring" reasons. You see, these "Aliens" I mean rich "Elites" are not dumb. They not only want to "mesmerize the minds of the masses" with all these various forms of Global Media we've been seeing throughout this study, but they also want to simultaneously know "which" of the masses around the world are not going to go along with their Subliminal Programming through these various forms of media. So, this is why they have devised a wicked scheme with Social Media that simultaneously not only allows them to rake in billions of dollars off of us, but to also literally catalog and profile any and all "resisters" who aren't going along with their "brainwashing" procedures through the Media. Especially those who even have the audacity to "put the glasses on" and "see" and then even "admit" or even "expose" what these "Elitists" are really up to on a global basis and then warn others of their nefarious plans.

This "monitoring" and "cataloging" of any "resisters" is not only going on whether we realize it or not but think of the "manipulative" irony going on here. We the individual who are being "monitored" by these Social Media companies to check for any resisters, have also been tricked into "volunteering" this "profiling" information to them, on us, for free! Pretty slick, isn't it? But as you can see, with the advances of Social Media in the last couple of decades, it's not only become a major Media staple to "mold our minds" and allow us to "receive information" in general. But just like Newspapers, Radio, Music, Books, the Education System, and Television, once this "manipulative" technology called Social Media sprang into existence, it too began to spread across the whole planet. And that's why today, the bulk of the planet is "connected" to all different kind of Social Media accounts.

"In May of 1997, SixDegrees was the first site to allow users to create profiles and friends lists. It was promoted as a tool to help people connect with and send messages to each other. SixDegrees closed in 2000.

Social Media site Friendster launches in 2002. Membership peeked at 115 million users in 2008, then steadily declined until closing in June of 2015. Inspired spin-offs included Dogster and Elfster.

MySpace, considered by many to be the first mainstream social media site, was acquired by News Corp in 2005 for $580 million. MySpace received more than 75 million visitors per month in late 2008.

Initially open to Harvard students in 2004, then opened to 800 colleges in May 2005. April 2008: Facebook's popularity overtakes MySpace's. In February 2009, the 'like' button debuts. Today, 68% of US adults use Facebook.

Corporate social networking site LinkedIn launched in 2003. In July 2011, LinkedIn became the #2 site in the US, surpassing MySpace's 33.5 million visitors with 33.9 million.

The first YouTube video was uploaded in April 2005, featuring co-founder Jawed Karim at the San Diego Zoo. Founded by three former employees of PayPal. The site has over 2 billion users, over 1/3 of all internet users! Every minute, more than 500 hours of video are uploaded to YouTube.

Text-based social media service Twitter was born in 2006. Messages called 'Tweets' were limited to 140 characters each. In August 2007 the hashtag was introduced. Today, approximately 350,000 tweets are posted every minute.

In 2010, Pinterest introduced unequal boxes, call 'pins,' for sharing information. Tutorials, guides and do-it-yourself pins have a 42% higher click rate versus all other types of pins. It is currently the fastest growing website by overall member growth at 57%.

Originally called 'Codename' before launch, the first photo ever posted (a stray dog) was by the co-founder, @kevin on July 16, 2010. A study found that 8% of all Instagram accounts are fake. The most 'Instagrammed' food is pizza, in front of sushi and steak.

Google Plus beta launched in 2011. After two weeks, more than 10 million users shared around 1B items per day. The site shut down in April 2019 due to low user engagement.

Pictures and messages sent are only accessible for a short time. Currently over 210 million active Snapchat users. 73% of US users are 18-24 years old. In the US, 90% of all 13–24-year-olds use the app.

More than 1 billion 'loops' were played daily in 2015 on the video sharing service. If you flip the Vine logo, it reveals the number 6, the length of the videos on Vine. Vine is short for 'vignette,' or a short impressionistic scene. The service shut down in January 2017, unable to keep up with competitor features.

TikTok began in September of 2016. The video-sharing social network received 738 million downloads in 2019. TikTok has over 800 million active users worldwide. The 2nd most popular free app download in 2019, behind WhatsApp. TikTok has the highest social engagement rates per post vs. Instagram & Twitter.[4]

Wow! 2.6 billion users on Facebook alone? That's a lot of global influence in the hands of just one "Alien" I mean rich "Elite" around the whole planet! And now, nearly 4 billion people currently use social media worldwide, which is an overall increase in users of 92.76% in just five years. And the average person even has around 9 social media accounts. It's worse than a Lay's potato chip apparently! You can't have just one! But as you can see, in all seriousness, the Social Media platforms are simply yet another fantastic leap forward of Manipulative Media Technology if you want to connect to the world and "mesmerize the minds of the masses" for some nefarious purpose! And that's exactly what these "Elites" are doing! Social Media has now become a very important, and frankly, "dependent" tool to share "information," "educate," or even "mold" the minds of people either for good or bad. Therefore, once again, whoever controls the Media of Social Media and all its various platforms, they have yet another powerful tool to "manipulate the minds of the masses" around the whole world. And believe it or not, that "control" of Social Media on a global basis is already here! Just like with Newspapers, Radio, Music, Books, the Education System, Television, so now even Social Media is in the hands of just a few "Aliens" I mean rich "Elites." You still seeing a pattern here?

This has already been demonstrated with the Top 10 Social Media owners that I shared with you that are raking in billions of dollars off their Social Media platforms. "They alone" control the bulk of the planet's Social Media outlets! And just like all the other forms of Global Media, so too are these handful of "Elite" Social Media owners using their global ownership of these platforms for their own evil "manipulative" purposes to regurgitate a "preprogrammed" "pre-scripted narrative" for "us" to digest from "them" to literally "program" our minds and "mesmerize the masses." Including the "silencing" of any "dissenting opinion" by people exercising their God-given "rights" to "freedom of speech." Again, the pattern is exactly the same as with all other forms of Media! These "Aliens" I mean rich "Elites" know what they're doing! And that horrible truth came out in the last election cycle when it was shown that these "Top Social Media" owners and their various platforms were "refusing" any "narrative" that they didn't "want" us to "see" or "hear" or even "listen" to involving the election process. In fact, "they" even "silenced" the President of the United States! I mean, who do "they" think they are? But this is what they did during and after the election.

Candace Owens: *"Facebook is trying to delete me! Big Tech is trying to delete us, our movement. We must stop them.*

You're smart. Long ago, you stopped believing in the corrupt corporate media. You canceled the newspapers and cable shows and began reading your news from social platforms. Dominant on those platforms were conservative voices like mine. We are the Americans who did not have a voice in traditional media. Honestly, do you think the left wants me to have a platform to speak? But I can have a voice, here on my own social platforms. (On her cell phone) I can be heard. That freedom infuriates the left, who used to own complete control over our information. So, they hatched a plan to silence me, and all of us conservatives, and Facebook is paying them for it. Here is how it works.

In 2016, hysterical liberals had to find someone to blame for their humiliating loss to Donald Trump. In their minds, they could not possibly have lost due to their own horrendous candidates or policies or their own failing message. So, they attacked the one thing that they did not have total control over, social media companies. They applied extreme pressure to silence or censor 'fake news' which was just a fancy way of saying news that they don't like.

Facebook bent to that pressure and created a fact-checker network with God-like powers over all of us. Here is how Facebook 'fact-checking' works. A website you have never heard of run by partisan beta leftists stalk the pages of your favorite conservative personalities. Whenever we say anything they disagree with, these fact-checkers write a vicious partisan hit piece. Then, they harass us and our audience by slapping hazardous warning labels on what we have posted. Many times, those labels say, 'missing context' or 'disputed.' Yes, thanks Facebook.

Every political argument in America is disputed. Every argument is indeed missing some context, but the insanity continues. These 'hack-checkers,' as I like to call them, would be just an annoyance, but Facebook gave them the ability to destroy the reach of our content so we cannot communicate with you. They de-monetize our pages so we can't make a living producing content for our audiences. And here's the breathtaking part. Most of the fact-checks are pure partisan bile. They're false, fake, untrue, but there is no recourse for your favorite conservative creators. We are held at gunpoint by these partisan hacks. And Facebook is giving them the gun.

This is why I am suing the Facebook fact-checkers. I am suing them on behalf of you, your favorite creators and news sites on behalf of our freedom of speech and thought. So, who are these hack-checkers? Well, I'm suing a site called Lead Stories. Lead Stories is run by a former CNN employee of 26 years, who makes a living cannibalizing and slandering conservative content and creators. His website has published disputed fake new stories, including one that went after my content personally. So, I'm going after him personally.

Here's an important question. Who funds Lead Stories? According to their own masthead, Lead Stories is funded by Facebook and Google, along with Bytedance. And what is Bytedance? It's a Beijing based company that owns the TikTok spyware app. Our own Justice Department says the company is a national security threat and compromised by the communist party of China. Facebook then, is paying for fact-checkers funded by an arm of the communist party of China to censor my content here in America.

The communist party of China has more say over my content than the First Amendment. This is insanity. It's un-American, it's dangerous and it needs to stop. This is why I am fighting. Our freedoms are at stake. Our movement is at stake. We know these battles are taking place all over the world."

Newscaster: *"This morning President Trump, waking up without his favorite megaphone."*

President Trump: *"Let me ask you. Should I keep the Twitter going or not?"*

Newscaster: *"Twitter permanently banning the Commander in Chief's personal account with 88 million followers. After an initial 12-hour lockout following Wednesday's riots at the US capitol. The Company run by Jack Dorsey saying, 'After close review of recent Tweets from the @realDonaldTrump account and the context around them we have* *permanently suspended the account due to the risk of further incitement of violence.'*

With his legend of tweets simply gone, President Trump moved quickly to post on the official POTUS account, attacking Twitter, Democrats, and the radical left. Those tweets were taken down within minutes. Twitter's actions sparking celebration and outrage. Senator Joe Manchin, thanking Twitter and Mark Warner calling it an overdue step.

But the President's son, Donald Trump, Jr. was furious, sounding off, 'Free-speech no longer exists in America, it died with big tech and what's left is only there for a chosen few.'

Trump's former UN Ambassador Nikki Haley saying, 'This is what happens in China, not our country.'"

Man in America: *"2020 has been a year like no other. Everything we know, everything we trust, and everything we came to rely on has changed. So how can we expect the US election to be any different? Mainstream media are telling us the race is over. Biden won! Time to move on. Nothing to see here folks.*

But in your gut, you know something's just not right. From the strange patterns on election night, to the weeks of suspense, to the reports of fraud that evaporate faster than Biden's memory. It just doesn't add up. Stay with me and you will see just what is going on behind the scenes in this election. And why the greatest threat we are facing right now is not the invisible enemy entering our bodies but one that is invading our country and our minds.

How do you know what is happening in the world right now? How do you know what is going on in this country? Or even your city? I'm going to go out on a limb here and say you probably turn to mainstream media, which mean social media. Perhaps you watch Tucker in the evenings or read the New York Times Sunday morning. Or you check Facebook at lunch on your break. So how do you know what is going on in the US election?

What did you think on election night when historically red states and battleground states were stalled while blue states were immediately called for Biden? Or when Fox called Arizona with so many votes still out. Of course, for Biden.

And how about the late-night ballot dumps that were somehow all for Biden? And the ballots that were pristine mail-in ballots that were all for Biden. Then there were the burst water pipes, and the power outages, and the computer glitches, and the issue of Sharpy pens, and lost USB drives that somehow all favored Biden. Then we have the hordes of dead voters. Once again, all for Biden. The dead people really did vote for Biden, I fact-checked it. And what are the odds that all the last-minute changes to the election process and voting laws in the name of Covid also favored Biden?

And how about all the counties with more votes than eligible voters or where they got more mail-in ballots that they even sent out? In Pennsylvania, they sent out 1.8 million ballots and got back 2.5 million. And last but not least we got Dominion, the voting system with ties to socialist Venezuela that was programmed to manipulate votes for Biden, while sending our data to China, Iran and Russia. When the system crashed because of the Trump landslide, key battleground states simply decided to stop counting and head on home.

When has this ever happened before in US history? This is just the tip of the iceberg. Every one of these reports should be serious grounds for investigation. Thousands of poll workers, postal workers, election officials, and honest Americans, both Republican and Democrat have come forward with photos, videos and firsthand stories, affirmed by sworn affidavits, documenting massive voter fraud. Social media had been flooded with them, yet mainstream media keeps asking, 'Where's the evidence?'"

NTDLive: *"Once everyone is gone, the coast is clear. They are going to pull ballots out from underneath a table. Watch this table. Do you see the gentleman in the red? He just pulled one out. (He pulled a box of ballots* *out from under the table.) So, what are these ballots doing there? Separate from all the other ballots? And why are they counting them when the place is cleared out? With no witnesses?"*

Man in America: *"Where is the evidence?"*

Witness: *"I'm here with my friend, and we were taking the trash out and I wanted to show you guys what we found in the trash cans. (He videos ballot after ballot that are all marked Donald Trump) There are more ballots in this trash bag and in the trash bin. But we will say, we did find one Biden."*

Man in America: *"Where is the evidence?"*

Rudy Giuliani: *"Can you calculate out how many voted for Biden and how much for Trump?"*

Col. Phil Waldron: *"Close to 600,000. I think our figures were about 570,000 some odd thousand, all those spikes represented over time."*

Rudy Giuliani: *"For Biden? And how much for Trump?""*

Col. Phil Waldron: *"I think it was a little over 3,200."*

Man in America: *"Now hold on! Didn't they just drag us through a baseless four-year witch hunt in the Russian election interference over far less evidence and isn't it their job to be digging for evidence? Would Nixon have gotten away with Watergate if the media didn't dig for evidence? Not only aren't they digging for it right now, but they are burying it in a giant game of Whack-a-Mole, along with their comrades over at Facebook, Twitter, and Google. Of course, they had their fact-checkers debunk everything first. These arbiters of truth, who have more confidence and interest in what flavors of ice cream are in Nancy Pelosi's freezer. Even President Trump's tweets are so-called fact-checked and censored, and his press conferences are interrupted and cut off. (As*

President Trump is speaking, the new commentator speaks over him making it impossible to understand what he is saying.)

News Commentator: *"I am not only interrupting the President of the United States but correcting the President of the United States."*

Man in America: *"Think about what this means. The President of the United States of America is being censored by US companies. This whole thing reeks of corruption. Not just corruption, but treason. It is such a joke that I would be laughing if it wasn't so serious. But it is serious! Because let's assume for a moment that just a fraction of the reports are true. If mainstream and social media didn't let you see or even talk about them, how would you even know? Think about it. If our election was stolen and our government was overthrown and yet they kept it all hidden, how would you know?*

If the mainstream media can completely control public opinion and the flow of information, including from our President, does the truth even matter? What if I told you that every single one of our mainstream media, including FOX is being used as a tool to manipulate public opinion and steer the outcome of the election? Perhaps a year ago you might have called me a Conspiracy Theorist but considering that we have been living in the Twilight Zone lately you are probably starting to realize that anything is possible.

Just try to wrap your head around this, how did the media so accurately predict the long delays in counting votes, in the red mirage before the blue wave. And why did they spend months downplaying the risks of mail-in ballots while sewing the seed that Trump would claim voter fraud and refuse to concede? Do they have a crystal ball? Why did Hillary Clinton tell Biden not to concede under any circumstances, all the way back in the summer? Why did they spend for the last four years, using every possible tactic, to undermine President Trumps presidency in the most relentless smear campaign that the world has ever seen?

And why did every mainstream media rush to coronate Biden and cement him in the public mind? Now do you understand why he told voters, 'I don't need you to get me elected.' It never mattered that he hid in his basement all summer. It never mattered that only 5 people turned out for his rallies. It didn't matter that he couldn't even string a sentence together. It didn't matter that the majority of Americans chose President Trump on November 3rd because when you control every single channel of information, nothing else matters.

Right now, big media, big tech, big government and even celebrities are blanketing us with the narrative that Biden won while censoring all evidence of fraud and concealing the very real truth that our nation has been facing its greatest threat since the American Revolution. Because as our Founding Fathers knew all too well, the moment we no longer have free speech and a free press to keep the government in check, we no longer have a democracy."[5]

And that's exactly what's happened from a handful of "Aliens" I mean rich "Elites" who now "own" and "control" all the "information" and "content" on all the various Global Media "platforms" including Social Media. Again, whatever happened to "free speech" including for the President of the United States?! Are you "Elites" actually "controlling" the "narrative" of what people "receive" from all these various Media outlets, including Social Media, to the point where you actually "silence" any dissenting view? Again, this is not news reporting, it's news determination! It's indoctrination and flat out brainwashing even in the United States of America! Our Founding Fathers must be rolling over in their graves! In fact, let me give you just one more example of just how much the information that we are receiving from the internet, as well as Social Media, really is "filtered" and "controlled" by a handful of "Aliens" I mean rich "Elites."

Let's take a look at another "manipulative entity called Google. I mean, surely when we use their services, including their search engine online, we're getting the truth, right? Just the facts, right? Wrong! They are actually "subliminally" allowing us to only "receive" what they "want" us to receive. Here's just some of the proof!

"Over the past century, more than a few great writers have expressed concern about humanity's future. In The Iron Heel (1908), the American writer Jack London pictured a world in which a handful of wealthy corporate titans – the 'oligarchs' – kept the masses at bay with a brutal combination of rewards and punishments. Much of humanity lived in virtual slavery, while the fortunate ones were bought off with decent wages that allowed them to live comfortably – but without any real control over their lives.

In We (1924), the brilliant Russian writer Yevgeny Zamyatin, anticipating the excesses of the emerging Soviet Union, envisioned a world in which people were kept in check through pervasive monitoring. The walls of their homes were made of clear glass, so everything they did could be observed. They were allowed to lower their shades an hour a day to have sex, but both the rendezvous time and the lover had to be registered first with the state.

In Brave New World (1932), the British author Aldous Huxley pictured a near-perfect society in which unhappiness and aggression had been engineered out of humanity through a combination of genetic engineering and psychological conditioning.

And in the much darker novel 1984 (1949), Huxley's compatriot George Orwell described a society in which thought itself was controlled; in Orwell's world, children were taught to use a simplified form of English called Newspeak in order to assure that they could never express ideas that were dangerous to society.

These are all fictional tales, to be sure, and in each, the leaders who held the power used conspicuous forms of control that at least a few people actively resisted and occasionally overcame. But in the non-fiction bestseller The Hidden Persuaders (1957) – recently released in a 50th-anniversary edition – the American journalist Vance Packard described a 'strange and rather exotic' type of influence that was rapidly emerging in the United States and that was, in a way, more threatening than the fictional types of control pictured in the novels.

According to Packard, US corporate executives and politicians were beginning to use subtle and, in many cases, completely undetectable methods to change people's thinking, emotions and behavior based on insights from psychiatry and the social sciences.

Most of us have heard of at least one of these methods: subliminal stimulation or what Packard called 'subthreshold effects' – the presentation of short messages that tell us what to do but that are flashed so briefly we aren't aware we have seen them.

In 1958, propelled by public concern about a theater in New Jersey that had supposedly hidden messages in a movie to increase ice cream sales, the National Association of Broadcasters – the association that set standards for US television – amended its code to prohibit the use of subliminal messages in broadcasting.

In 1974, the Federal Communications Commission opined that the use of such messages was 'contrary to the public interest.' Legislation to prohibit subliminal messaging was also introduced in the US Congress but never enacted. Both the UK and Australia have strict laws prohibiting it. Subliminal stimulation is probably still in wide use in the US – it's hard to detect, after all, and no one is keeping track of it.

Packard had uncovered a much bigger problem, however – namely that powerful corporations were constantly looking for, and in many cases already applying, a wide variety of techniques for controlling people without their knowledge. He described a kind of cabal in which marketers worked closely with social scientists to determine, among other things, how to get people to buy things they didn't need and how to condition young children to be good consumers – inclinations that were explicitly nurtured and trained in Huxley's Brave New World.

Guided by social science, marketers were quickly learning how to play upon people's insecurities, frailties, unconscious fears, aggressive feelings and sexual desires to alter their thinking, emotions and behavior without any awareness that they were being manipulated.

By the early 1950s, Packard said, politicians had got the message and were beginning to merchandise themselves using the same subtle forces being used to sell soap. Packard prefaced his chapter on politics with an unsettling quote from the British economist Kenneth Boulding: 'A world of unseen dictatorship is conceivable, still using the forms of democratic government.' Could this really happen, and, if so, how would it work?

The forces that Packard described have become more pervasive over the decades. The soothing music we all hear overhead in supermarkets causes us to walk more slowly and buy more food, whether we need it or not.

Most of the vacuous thoughts and intense feelings our teenagers experience from morning till night are carefully orchestrated by highly skilled marketing professionals working in our fashion and entertainment industries.

Politicians work with a wide range of consultants who test every aspect of what the politicians do in order to sway voters: clothing, intonations, facial expressions, makeup, hairstyles and speeches are all optimized, just like the packaging of a breakfast cereal.

But what would happen if new sources of control began to emerge that had little or no competition? And what if new means of control were developed that were far more powerful – and far more invisible – than any that have existed in the past? And what if new types of control allowed a handful of people to exert enormous influence not just over the citizens of the US but over most of the people on Earth?

It might surprise you to hear this, but these things have already happened.

Google decides which web pages to include in search results, and how to rank them. How it does so is one of the best-kept secrets in the world, like the formula for Coca-Cola.

To understand how the new forms of mind control work, we need to start by looking at the search engine – one in particular: the biggest and best of them all, namely Google. The Google search engine is so good and so popular that the company's name is now a commonly used verb in languages around the world. To 'Google' something is to look it up on the Google search engine, and that, in fact, is how most computer users worldwide get most of their information about just about everything these days. They Google it.

Google has become the main gateway to virtually all knowledge, mainly because the search engine is so good at giving us exactly the information we are looking for, almost instantly, and almost always in the first position of the list it shows us after we launch our search – the list of 'search results.'

That ordered list is so good, in fact, that about 50 per cent of our clicks go to the top two items, and more than 90 per cent of our clicks go to the 10 items listed on the first page of results; few people look at other results pages, even though they often number in the thousands, which means they probably contain lots of good information.

Google decides which of the billions of web pages it is going to include in our search results, and it also decides how to rank them. How it decides these things is a deep, dark secret – one of the best-kept secrets in the world, like the formula for Coca-Cola.

Because people are far more likely to read and click on higher-ranked items, companies now spend billions of dollars every year trying to trick Google's search algorithm – the computer program that does the selecting and ranking – into boosting them another notch or two. Moving up a notch can mean the difference between success and failure for a business and moving into the top slots can be the key to fat profits."

Again, there's your "monetary" manipulation to the tune of billions, or in this case "fat profits." But do you see how the whole system is "rigged" and "controlled" whether you realized it or not? It's just another long line of "subliminal seduction" foisted upon us for various nefarious purposes by these global "Elites!" In fact, it has now come out that Google has not only been guilty for a long time of "subliminally censoring" content for "manipulative" "monetary" purposes, but they and other Social Media platforms are also simultaneously "blacklisting" entities and people who are trying to share something contrary to what Google "wants" us to hear. Again, whatever happened to "free speech?" But check this proof out.

"The New Censorship. How did Google become the Internet's censor and master manipulator, blocking access to millions of websites? Google, Inc., isn't just the world's biggest purveyor of information; it is also the world's biggest censor.

The company maintains at least nine different blacklists that impact our lives, generally without input or authority from any outside advisory group, industry association or government agency.

Google is not the only company suppressing content on the internet. Reddit has frequently been accused of banning postings on specific topics, and a recent report suggests that Facebook has been deleting conservative news stories from its newsfeed, a practice that might have a significant effect on public opinion – even on voting. Google, though, is currently the biggest bully on the block.

When Google's employees or algorithms decide to block our access to information about a news item, political candidate or business, opinions and votes can shift, reputations can be ruined, and businesses can crash and burn. Because online censorship is entirely unregulated at the moment, victims have little or no recourse when they have been harmed.

Should we allow a large corporation to wield an especially destructive kind of power that should be exercised with great restraint and should belong only to the public: the power to shame or exclude?

If Google were just another mom-and-pop shop with a sign saying, 'we reserve the right to refuse service to anyone,' that would be one thing. But as the golden gateway to all knowledge, Google has rapidly become an essential in people's lives – nearly as essential as air or water. We don't let public utilities make arbitrary and secretive decisions about denying people services; we shouldn't let Google do so either.

Let's start with the most trivial blacklist and work our way up. I'll save the biggest and baddest – one the public knows virtually nothing about but that gives Google an almost obscene amount of power over our economic well-being – until last.

1. The autocomplete blacklist. *This is a list of words and phrases that are excluded from the autocomplete feature in Google's search bar. The search bar instantly suggests multiple search options when you type words such as 'democracy' or 'watermelon,' but it freezes when you type 'other' certain words. The autocomplete blacklist can also be used to protect or discredit political candidates. As recently reported, at the moment autocomplete shows you 'Ted' (for former GOP presidential candidate Ted Cruz) when you type 'lying,' but it will not show you 'Hillary' when you type 'crooked' – not even, on my computer anyway, when you type 'crooked hill.' (The nicknames for Clinton and Cruz coined by Donald Trump, of course.) If you add the 'a,' so you've got 'crooked hilla,' you get the very odd suggestion 'crooked Hillary Bernie.' When you type 'crooked' on Bing, 'crooked Hillary' pops up instantly. Google's list of forbidden terms varies by region and individual.*

2. The Google Maps blacklist. *This list is a little creepier, especially if you are concerned about your privacy. The cameras of Google Earth and Google Maps have photographed your home for all to see. If you don't like that, 'just move,' Google's former CEO Eric Schmidt said. Google also maintains a list of properties it either blacks out or blurs out in its* images. Some are probably military installations, some the residences of wealthy people, and some – well, who knows? Martian pre-invasion enclaves? Google doesn't say.

3. The YouTube blacklist. YouTube, which is owned by Google, allows users to flag inappropriate videos, at which point Google censors weigh in and sometimes remove them, but not, according to a recent report by Gizmodo, with any great consistency – except perhaps when it comes to politics. Consistent with the company's strong and open support for liberal political candidates, Google employees seem far more apt to ban politically conservative videos than liberal ones. In December 2015, singer Joyce Bartholomew sued YouTube for removing her openly pro-life music video. YouTube also sometimes acquiesces to the censorship demands of foreign governments. Most recently, in return for overturning a three-year ban on YouTube in Pakistan, it agreed to allow Pakistan's government to determine which videos it can and cannot post.

4. The Google account blacklist. A couple of years ago, Google consolidated a number of its products – Gmail, Google Docs, Google+, YouTube, Google Wallet and others – so you can access all of them through your one Google account. If you somehow violate Google's vague and intimidating terms of service agreement, you will join the ever-growing list of people who are shut out of their accounts, which means you'll lose access to all of these interconnected products. Because virtually no one has ever read this lengthy, legalistic agreement however, people are shocked when they're shut out, in part because Google reserves the right to 'stop providing Services to you…at any time.' And because Google, one of the largest and richest companies in the world, has no customer service department, getting reinstated can be difficult.

5. The Google News blacklist. If a librarian were caught trashing all the liberal newspapers before people could read them, he or she might get in a heap o' trouble. What happens when most of the librarians in the world have been replaced by a single company? Google is now the largest news aggregator in the world, tracking tens of thousands of news sources in more than thirty languages and recently adding thousands of small, local news sources to its inventory. It also selectively bans news sources as it pleases. In 2006, Google was accused of excluding conservative news sources that generated stories critical of Islam, and the company has also been accused of banning individual columnists and competing companies from its news feed. In December 2014, facing a new law in Spain that would have charged Google for scraping content from Spanish news sources (which, after all, have to pay to prepare their news), Google suddenly withdrew its news service from Spain, which led to an immediate drop in traffic to Spanish news stories. That drop in traffic is the problem: When a large aggregator bans you from its service, fewer people find your news stories, which means opinions will shift away from those you support. Selective blacklisting of news sources is a powerful way of promoting a political, religious or moral agenda, with no one the wiser.

6. The Google AdWords blacklist. Now things get creepier. More than 70 percent of Google's $80 billion in annual revenue comes from its AdWords advertising service, which it implemented in 2000 by infringing on a similar system already patented by Overture Services. The way it works is simple: Businesses worldwide bid on the right to use certain keywords in short text ads that link to their websites (those text ads are the AdWords); when people click on the links, those businesses pay Google. These ads appear on Google.com and other Google websites and are also interwoven into the content of more than a million non-Google websites – Google's 'Display Network.' The problem here is that if a Google executive decides your business or industry doesn't meet its moral standards, it bans you from AdWords; these days, with Google's reach so large, that can quickly put you out of business.

7. The Google AdSense blacklist. If your website has been approved by AdWords, you are eligible to sign up for Google AdSense, a system in which Google places ads for various products and services on your website. When people click on those ads, Google pays you. If you are good at driving traffic to your website, you can make millions of dollars a year running AdSense ads – all without having any products or services of your own. Meanwhile, Google makes a net profit by charging the companies behind the ads for bringing them customers; this accounts for about 18 percent of Google's income. Here, too, there is scandal: In April 2014, in two posts on PasteBin.com, someone claiming to be a former Google employee working in their AdSense department alleged the department engaged in a regular practice of dumping AdSense customers just before Google was scheduled to pay them, in some cases payments as high as $500,000.

8. The search engine blacklist. Google's ubiquitous search engine has indeed become the gateway to virtually all information, handling 90 percent of search in most countries. It dominates search because its index is so large: Google indexes more than 45 billion web pages; its next-biggest competitor, Microsoft's Bing, indexes a mere 14 billion, which helps to explain the poor quality of Bing's search results. Google's dominance in search is why businesses large and small live-in constant 'fear of Google,' as Mathias Dopfner, CEO of Axel Springer, the largest publishing conglomerate in Europe, put it in an open letter to Eric Schmidt in 2014. According to Dopfner, when Google made one of its frequent adjustments to its search algorithm, one of his company's subsidiaries dropped dramatically in the search rankings and lost 70 percent of its traffic within a few days. Even worse than the vagaries of the adjustments, however, are the dire consequences that follow when Google employees somehow conclude you have violated their 'guidelines:' You either get banished to the rarely visited Netherlands of search pages beyond the first page (90 percent of all clicks go to links on that first page) or completely removed from the index. Search ranking manipulations of this sort don't just ruin businesses; they also affect people's opinions, attitudes, beliefs and behavior, as my research on the Search Engine Manipulation Effect has demonstrated. This brings me, at last, to the biggest and potentially most dangerous of Google's blacklists – which Google's calls its 'quarantine' list.

9. The quarantine list. To get a sense of the scale of this list, I find it helpful to think about an old movie – the classic 1951 film "The Day the Earth Stood Still," which starred a huge metal robot named Gort. He had *laser-weapon eyes, zapped terrified humans into oblivion and had the power to destroy the world. Klaatu, Gort's alien master, was trying to deliver an important message to earthlings, but they kept shooting him before he could. Finally, to get the world's attention, Klaatu demonstrated the enormous power of the alien races he represented by shutting down – at noon New York time – all of the electricity on earth for exactly 30 minutes. The earth stood still. Substitute 'ogle' for 'rt,' and you get 'Google,' which is every bit as powerful as Gort – so good, in fact, that you are probably unaware that on Jan. 31, 2009, Google blocked access to virtually the entire internet. And, as if not to be outdone by a 1951 science fiction move, it did so for 40 minutes.*

Impossible, you say. Why would do-no-evil Google do such an apocalyptic thing, and, for that matter, how, technically, could a single company block access to more than 100 million websites? The answer has to do with the dark and murky world of website blacklists – ever-changing lists of websites that contain malicious software that might infect or damage people's computers. In 2012, Google acknowledged that each and every day it adds about 9,500 new websites to its quarantine list and displays malware warnings on the answers it gives to between 12 and 14 million search queries. It won't reveal the exact number of websites on the list, but it is certainly in the millions on any given day. In 2011, Google blocked an entire subdomain, co.cc, which alone contained 11 million websites, justifying its action by claiming that most of the websites in that domain appeared to be 'spammy.' According to Matt Cutts, still the leader of Google's web spam team, the company 'reserves the right' to take such action when it deems it necessary. (The right? Who gave Google that right?)

And that's nothing: According to The Guardian, on Saturday, Jan. 31, 2009, at 2:40 pm GMT, Google blocked the entire internet for those impressive 40 minutes, supposedly, said the company, because of 'human error' by its employees. It would have been 6:40 am in Mountain View, California, where Google is headquartered. Was this time chosen because it is one of the few hours of the week when all of the world's stock markets are closed?

When Google's search engine shows you a search result for a site it has quarantined, you see warnings such as, 'The site ahead contains malware' or 'This site may harm your computer' on the search result. That's useful information if that website actually contains malware, either because the website was set up by bad guys or because a legitimate site was infected with malware by hackers. But Google's crawlers often make mistakes, blacklisting websites that have merely been 'hijacked,' which means the website itself isn't dangerous but merely that accessing it through the search engine will forward you to a malicious site. My own website was hijacked in this way in early 2012. Accessing the website directly wasn't dangerous but trying to access it through the Google search engine forwarded users to a malicious website in Nigeria. When this happens, Google not only warns you about the infected website on its search engine (which makes sense), it also blocks you from accessing the website directly through multiple browsers – even non-Google browsers.

The mistakes are just one problem. The bigger problem is that even though it takes only a fraction of a second for a crawler to list you, after your site has been cleaned up Google's crawlers sometimes take days or even weeks to delist you – long enough to threaten the existence of some businesses. This is quite bizarre considering how rapidly automated online systems operate these days. Within seconds after you pay for a plane ticket online, your seat is booked, your credit card is charged, your receipt is displayed, and a confirmation email shows up in your inbox – a complex series of events involving multiple computers controlled by at least three or four separate companies. But when you inform Google's automated blacklist system that your website is now clean, you are simply advised to check back occasionally to see if any action has been taken. To get delisted after your website has been repaired, you either have to struggle with the company's online Webmaster tools, which are far from friendly, or you have to hire a security expert to do so – typically for a fee ranging between $1,000 and $10,000. No expert, however, can speed up the mysterious delisting process; the best he or she can do is set it in motion.

So far, all I've told you is that Google's crawlers scan the internet, sometimes find what appear to be suspect websites and put those websites on a quarantine list. That information is then conveyed to users through the search engine. So far so good, except of course for the mistakes and the delisting problem; one might even say that Google is performing a public service, which is how some people who are familiar with the quarantine list defend it. But I also mentioned that Google somehow blocks people from accessing websites directly through multiple browsers. How on earth could it do that? How could Google block you when you are trying to access a website using Safari, an Apple product, or Firefox, a browser maintained by Mozilla, the self-proclaimed 'nonprofit defender of the free and open internet?'

The key here is browsers. No browser maker wants to send you to a malicious website, and because Google has the best blacklist, major browsers such as Safari and Firefox – and Chrome, of course, Google's own browser, as well as browsers that load through Android, Google's mobile operating system – check Google's quarantine list before they send you to a website.

If the site has been quarantined by Google, you see one of those big, scary images that say things like 'Get me out of here!' or 'Reported attack site!' At this point, given the default security settings on most browsers, most people will find it impossible to visit the site – but who would want to? If the site is not on Google's quarantine list, you are sent on your way.

OK, that explains how Google blocks you even when you're using a non-Google browser, but why do they block you? In other words, how does blocking you feed the ravenous advertising machine – the sine qua non of Google's existence?

Have you figured it out yet? The scam is as simple as it is brilliant: When a browser queries Google's quarantine list, it has just shared information with Google. With Chrome and Android, you are always giving up information to Google, but you are also doing so even if you are using non-Google browsers. That is where the money is – more information about search activity kindly provided by competing browser companies. How much information is shared will depend on the particular deal the browser company has with Google. In a maximum information deal, Google will learn the identity of the user; in a minimum information deal, Google will still learn which websites people want to visit – valuable data when one is in the business of ranking websites. Google can also charge fees for access to its quarantine list, of course, but that's not where the real gold is.

Chrome, Android, Firefox and Safari currently carry about 92 percent of all browser traffic in the U.S. – 74 percent worldwide – and these numbers are increasing. As of this writing, that means Google is regularly collecting information through its quarantine list from more than 2.5 billion people. Given the recent pact between Microsoft and Google, in coming months we might learn that Microsoft – both to save money and to improve its services – has also started using Google's quarantine list in place of its own much smaller list; this would further increase the volume of information Google is receiving.

To put this another way, Google has grown, and is still growing, on the backs of some of its competitors, with end users oblivious to Google's antics – as usual. It is yet another example of what I have called 'Google's Dance' – the remarkable way in which Google puts a false and friendly public face on activities that serve only one purpose for the company: increasing profit. On the surface, Google's quarantine list is yet another way Google helps us, free of charge, breeze through our day safe and well-informed. Beneath the surface, that list is yet another way Google accumulates more information about us to sell to advertisers.

You may disagree, but in my view, Google's blacklisting practices put the company into the role of thuggish internet cop – a role that was never authorized by any government, nonprofit organization or industry association. It is as if the biggest bully in town suddenly put on a badge and started patrolling, shuttering businesses as it pleased, while also secretly peeping into windows, taking photos and selling them to the highest bidder.

Consider: Heading into the holiday season in late 2013, an online handbag business suffered a 50 percent drop in business because of blacklisting. In 2009, it took an eco-friendly pest control company 60 days to leap the hurdles required to remove Google's warnings, long enough to nearly go broke. And sometimes the blacklisting process appears to be personal: In May 2013, the highly opinionated PC Magazine columnist John Dvorak wondered 'When Did Google Become the Internet Police?' after both his website and podcast site were blacklisted. He also ran into the delisting problem: 'It's funny,' he wrote, 'how the site can be blacklisted in a millisecond by an analysis, but I have to wait forever to get cleared by the same analysis doing the same scan. Why is that?'

How frequently Google acts irresponsibly is beside the point; it has the ability to do so, which means that in a matter of seconds any of Google's 37,000 employees with the right passwords or skills could laser a business or political candidate into oblivion or even freeze much of the world's economy."

Wow! That's a lot of "restrictive" "manipulative" power in the hands of just one "Alien" I mean rich "Elite" organization across the whole planet! And speaking of "severe global restrictive censorship" by Google and other various Social Media giants like Facebook and Twitter, we also saw this nefarious behavior when it came to sharing, or should we say "attempted" sharing, of information concerning Covid-19. Again, even if a person doesn't agree with the establishment's views, behavior, or action in regard to the Covid 19 virus and the handling of it, shouldn't a person still have the 'right" to "free speech" you know to say what they want to say about it? I mean, after all, this is the United States of America, right? The land of the "free." Well apparently, not anymore. See if this "censorship" sounds familiar!

"In February, when the new coronavirus began to spread outside China, the Director General of the World Health Organization (WHO), Tedros Adhanom Ghebreyesus, announced: 'we're not just fighting an epidemic; we're fighting an infodemic.' The term, coined in 2003 in the context of

the first SARS epidemic, refers to a rapid proliferation of information that is often false or uncertain.

Academic researchers, international organizations such as the United Nations and the European Union, individual governments and the media have acknowledged and discussed the prevalence of the alleged COVID-19 infodemic and the importance of fighting it.

Information campaigns have been launched to provide wider audiences with reliable information about COVID-19. Main social media platforms have also actively fought against false information by filtering out or flagging content considered as misinformation.

Censorship on major social media platforms, such as Facebook, Twitter and YouTube, is not a new phenomenon. These companies regularly remove content that they consider as objectionable based on continually updated categories outlined in their policies. Examples of 'objectionable content' include 'hate speech,' 'glorification of violence' or 'harmful and dangerous content.'

These categories are not only often broader than the exceptions to the freedom of speech entrenched in legislations of democratic countries, but also implicitly vague and leave plenty of room to interpretation. Indeed, an analysis of content banned on social networks suggests that the moderation is often politically biased.

Some very recent examples of moderation with apparent political ramifications include Twitter's labeling of US President Donald J. Trump's tweets as violating Twitter's policy about glorifying violence or abusive behavior, or adding a warning suggesting that a post was factually inaccurate.

Social media platforms are private companies and as such, one could argue, they should be able to decide what content they tolerate or not. However, such a view overlooks salient aspects of the issue. First, censorship on Facebook, Twitter or YouTube appears to contradict the very idea of these communication networks, that is, of spaces where everyone can express their opinion.

YouTube, for example, declares on its website that its 'mission is to give everyone a voice.' Twitter's manager once described his company as 'the free speech wing of the free speech party.' Mark Zuckerberg, the CEO of Facebook, has similarly been vocal about Facebook's commitment to the freedom of speech.

Many users of social media might have believed in these ideals when joining the online communities. Or, at least, they did not expect biased censorship on the platforms. From this point of view, appeals to the freedom of speech made by the social networks seem unfair or deceptive.

Another important point to consider is the fact that a few big tech companies currently dominate social media services, which also serve as a source of news to many users. According to a 2020 survey by the Reuters Institute for the Study of Journalism, 36% of 24,000 respondents from 12 countries use Facebook for news weekly, while 21% of the surveyed use YouTube for the same purpose.

If we add to this the fact that Google is the most popular search engine, it becomes clear that a few tech companies have huge power over what information Internet users can see and how their views are shaped.

What is clear, is that the tech giants' role in shaping public discourse has become apparent. The governments of various countries have either attempted to find a solution to this issue or to use the possibility of censorship by big tech companies for their own purposes.

While it is difficult to overlook the politically motivated censorship on online platforms and its implications, the removal of misinformation related to medical topics such as COVID-19 may seem to belong to a different category—not political, but rather one of science, where information can be objectively judged based on scientific evidence. At a closer look, however, this does not seem to be the case.

Censorship of information about COVID-19. In response to calls to combat misinformation about COVID-19, a group of companies, including among others Facebook, Twitter and YouTube, issued a joint statement in mid-March this year. They stated that they are 'jointly combating fraud and misinformation about the virus, elevating authoritative content on [their] platforms.'

Their actions include the introduction of 'educational pop-ups connecting people to information from the WHO' (Facebook), adding warning labels to content considered as false or misleading (Facebook, Twitter), removing content contradicting health authorities or the WHO (YouTube) and content that could directly contribute or lead to (physical) harm (Facebook and Twitter).

One example is a video removed by YouTube, in which a researcher, John Ioannidis, discussed data related to COVID-19, questioned the need to continue the ongoing lockdown and raised concerns about the negative impact of the restrictions. Other cases of censorship on major social media platforms have been reported, for example, removal of information about anti-quarantine protests on Facebook.

A major question regarding the policies of the communication platforms is who exactly defines and how which information is deemed to be false or harmful? And can we rely on these judgments?

One of the authoritative sources that all three major social media platforms mention in their policies on COVID-19 is the WHO. It is an established and influential organization, yet it may make mistakes, including in the context of handling epidemics. For example, concerns have been raised about influences of pharmaceutical companies on the guidelines related to the flu pandemic in 2009. A major question regarding the policies of the communication platforms is who exactly defines and how which information is deemed to be false or harmful?

YouTube and Twitter also refer to guidelines from local health authorities. Although these are usually developed by experts and may be legally binding, this does not imply that they are unerring. There has been disagreement among researchers in medical sciences about the necessity for lockdown measures. Furthermore, researchers and many healthcare professionals have indicated numerous and serious negative impacts of the policies introduced to combat the spread of COVID-19, and expressed doubts about the evidence supporting these measures.

This variety of opinions on how to handle the COVID-19 pandemic is related, among others, to the fact that it is a new disease and the knowledge about it is relatively limited and unsettled. Moreover, the implications of the pandemic and measures taken to counteract it exceed the remit of epidemiology or public health experts and fall into areas of economy, education, psychology and sociology. Meanwhile, experts who develop policies or express opinions about COVID-19 may not have a complete overview of the implications of pandemic-related policies. Processes of reviewing research results, drawing conclusions, and preparing guidelines may be complex, prone to mistakes and not immune to political or commercial interests.

Additionally, there are the 'usual' problems related to evaluation and translation of evidence into medical or public health practice. They include questions about the validity of a given study, limitations of methods, reproducibility of results and so on. Processes of reviewing research results, drawing conclusions and preparing guidelines may be complex, prone to mistakes and not immune to political or commercial interests. Retracted articles on COVID-19, including publications in The Lancet and the New England Journal of Medicine, suggest that research on COVID-19 is not an exception to problems related to the ethics of research.

Constructive critique, questioning of evidence and opinions of scientists and policymakers are thus necessary to identify and correct potential errors and to prevent them from being propagated. By following their policies on COVID-19, social media platforms filter out content which contradicts specific views that are not necessarily correct or unanimously accepted, with respect to the underlying scientific evidence or represented values and political views.

If critique of these views is eliminated or restricted, the possibility to correct errors, contribute to the understanding of the topic and informed public debate is limited. Additionally, since the censorship is not based solely on science – as scientific evidence is currently limited and medical experts still disagree on various topics – other factors influence decisions to remove content.

Questions about the commitment to the freedom of speech of the social media providers and risk of manipulation of public opinion are therefore relevant also in case of information about COVID-19."

I couldn't have put it better myself! Notice how they too use the word "manipulation" concerning public opinion! This is going on folks, on a massive global scale by these Tech and Social Media companies, whether we realize it or not! Even secular researchers are admitting it! And speaking of "manipulation," as if what we've seen so far wasn't bad enough, it is now known that Google and the various Social Media giants are not only "subliminally" "controlling" and "filtering" what information we receive from them, but they are now guilty of even "subliminally" influencing "election results" even here in the United States! I know it sounds crazy, but here's more of that proof!

"Late in 2012, I began to wonder whether highly ranked search results could be impacting more than consumer choices. Perhaps, I speculated, a top search result could have a small impact on people's opinions about things.

Early in 2013, with my associate Ronald E. Robertson of the American Institute for Behavioral Research and Technology in Vista, California, I put this idea to a test by conducting an experiment in which 102 people from the San Diego area were randomly assigned to one of three groups.

In one group, people saw search results that favored one political candidate – that is, results that linked to web pages that made this candidate look better than his or her opponent. In a second group, people saw search rankings that favored the opposing candidate, and in the third group – the control group – people saw a mix of rankings that favored neither candidate. The same search results and web pages were used in each group; the only thing that differed for the three groups was the ordering of the search results.

To make our experiment realistic, we used real search results that linked to real web pages. We also used a real election – the 2010 election for the prime minister of Australia. We used a foreign election to make sure that our participants were 'undecided'. Their lack of familiarity with the candidates assured this. Through advertisements, we also recruited an ethnically diverse group of registered voters over a wide age range in order to match key demographic characteristics of the US voting population.

All participants were first given brief descriptions of the candidates and then asked to rate them in various ways, as well as to indicate which candidate they would vote for; as you might expect, participants initially favored neither candidate on any of the five measures we used, and the vote was evenly split in all three groups. Then the participants were given up to 15 minutes in which to conduct an online search using 'Kadoodle', our mock search engine, which gave them access to five pages of search results that linked to web pages. People could move freely between search results and web pages, just as we do when using Google. When participants completed their search, we asked them to rate the candidates again, and we also asked them again who they would vote for.

We predicted that the opinions and voting preferences of 2 or 3 per cent of the people in the two bias groups – the groups in which people were seeing rankings favoring one candidate – would shift toward that candidate. What we actually found was astonishing. The proportion of people favoring the search engine's top-ranked candidate increased by 48.4 per cent, and all five of our measures shifted toward that candidate. What's more, 75 percent of the people in the bias groups seemed to have been completely unaware that they were viewing biased search rankings. In the control group, opinions did not shift significantly.

This seemed to be a major discovery. The shift we had produced, which we called the Search Engine Manipulation Effect (or SEME, pronounced 'seem'), appeared to be one of the largest behavioral effects ever discovered. We did not immediately uncork the Champagne bottle, however. For one thing, we had tested only a small number of people, and they were all from the San Diego area.

Over the next year or so, we replicated our findings three more times, and the third time was with a sample of more than 2,000 people from all 50 US states. In that experiment, the shift in voting preferences was 37.1 per cent and even higher in some demographic groups – as high as 80 per cent, in fact.

We also learned in this series of experiments that by reducing the bias just slightly on the first page of search results – specifically, by including one search item that favored the other candidate in the third or fourth position of the results – we could mask our manipulation so that few or even no people were aware that they were seeing biased rankings. We could still produce dramatic shifts in voting preferences, but we could do so invisibly.

Still no Champagne, though. Our results were strong and consistent, but our experiments all involved a foreign election – that 2010 election in Australia. Could voting preferences be shifted with real voters in the middle of a real campaign? We were skeptical. In real elections, people are bombarded with multiple sources of information, and they also know a lot about the candidates. It seemed unlikely that a single experience on a search engine would have much impact on their voting preferences.

To find out, in early 2014, we went to India just before voting began in the largest democratic election in the world – the Lok Sabha election for prime minister. The three main candidates were Rahul Gandhi, Arvind

Kejriwal, and Narendra Modi. Making use of online subject pools and both online and print advertisements, we recruited 2,150 people from 27 of India's 35 states and territories to participate in our experiment. To take part, they had to be registered voters who had not yet voted and who were still undecided about how they would vote.

Participants were randomly assigned to three search-engine groups, favoring, respectively, Gandhi, Kejriwal or Modi. As one might expect, familiarity levels with the candidates was high – between 7.7 and 8.5 on a scale of 10. We predicted that our manipulation would produce a very small effect, if any, but that's not what we found. On average, we were able to shift the proportion of people favoring any given candidate by more than 20 per cent overall and more than 60 per cent in some demographic groups. Even more disturbing, 99.5 per cent of our participants showed no awareness that they were viewing biased search rankings – in other words, that they were being manipulated.

SEME's near invisibility is curious indeed. It means that when people – including you and me – are looking at biased search rankings, they look just fine. So, if right now you Google 'US presidential candidates', the search results you see will probably look fairly random, even if they happen to favor one candidate. Even I have trouble detecting bias in search rankings that I know to be biased (because they were prepared by my staff). Yet our randomized, controlled experiments tell us over and over again that when higher-ranked items connect with web pages that favor one candidate, this has a dramatic impact on the opinions of undecided voters, in large part for the simple reason that people tend to click only on higher-ranked items. This is truly scary: like subliminal stimuli, SEME is a force you can't see; but unlike subliminal stimuli, it has an enormous impact – like Casper the ghost pushing you down a flight of stairs.

We published a detailed report about our first five experiments on SEME in the prestigious Proceedings of the National Academy of Sciences (PNAS) in August 2015. We had indeed found something important, especially given Google's dominance over search. Google has a near-

monopoly on internet searches in the US, with 83 per cent of Americans specifying Google as the search engine they use most often, according to the Pew Research Center. So, if Google favors one candidate in an election, its impact on undecided voters could easily decide the election's outcome.

Keep in mind that we had had only one shot at our participants. What would be the impact of favoring one candidate in searches people are conducting over a period of weeks or months before an election? It would almost certainly be much larger than what we were seeing in our experiments.

Other types of influence during an election campaign are balanced by competing sources of influence – a wide variety of newspapers, radio shows and television networks, for example – but Google, for all intents and purposes, has no competition, and people trust its search results implicitly, assuming that the company's mysterious search algorithm is entirely objective and unbiased. This high level of trust, combined with the lack of competition, puts Google in a unique position to impact elections. Even more disturbing, the search-ranking business is entirely unregulated, so Google could favor any candidate it likes without violating any laws. Some courts have even ruled that Google's right to rank-order search results as it pleases is protected as a form of free speech.

Does the company ever favor particular candidates? In the 2012 US presidential election, Google and its top executives donated more than $800,000 to President Barack Obama and just $37,000 to his opponent, Mitt Romney. And in 2015, a team of researchers from the University of Maryland and elsewhere showed that Google's search results routinely favored Democratic candidates. Are Google's search rankings really biased? An internal report issued by the US Federal Trade Commission in 2012 concluded that Google's search rankings routinely put Google's financial interests ahead of those of their competitors, and anti-trust actions currently under way against Google in both the European Union and India are based on similar findings.

In most countries, 90 percent of online search is conducted on Google, which gives the company even more power to flip elections than it has in the US and, with internet penetration increasing rapidly worldwide, this power is growing. In our PNAS article, Robertson and I calculated that Google now has the power to flip upwards of 25 percent of the national elections in the world with no one knowing this is occurring. In fact, we estimate that, with or without deliberate planning on the part of company executives, Google's search rankings have been impacting elections for years, with growing impact each year. And because search rankings are ephemeral, they leave no paper trail, which gives the company complete deniability.

Power on this scale and with this level of invisibility is unprecedented in human history. But it turns out that our discovery about SEME was just the tip of a very large iceberg.

Recent reports suggest that former Democratic presidential candidate Hillary Clinton made heavy use of social media to try to generate support – Twitter, Instagram, Pinterest, Snapchat and Facebook, for starters. She had 5.4 million followers on Twitter, and her staff was tweeting several times an hour during waking hours.

Is social media as big a threat to democracy as search rankings appear to be? What happens if such technologies are misused by the companies that own them? A study by Robert M Bond, now a political science professor at Ohio State University, and others published in Nature in 2012 described an ethically questionable experiment in which, on election day in 2010, Facebook sent 'go out and vote' reminders to more than 60 million of its users. The reminders caused about 340,000 people to vote who otherwise would not have.

Writing in the New Republic in 2014, Jonathan Zittrain, professor of international law at Harvard University, pointed out that, given the massive amount of information it has collected about its users, Facebook could easily send such messages only to people who support one particular party or candidate, and that doing so could easily flip a close election – with no one knowing that this has occurred. And because advertisements, like search rankings, are ephemeral, manipulating an election in this way would leave no paper trail.

Are there laws prohibiting Facebook from sending out ads selectively to certain users? Absolutely not, in fact, targeted advertising is how Facebook makes its money. Is Facebook currently manipulating elections in this way? No one knows, but some candidates are better for a company than others, and Facebook's executives have a fiduciary responsibility to the company's stockholders to promote the company's interests.

Another Facebook experiment, published in 2014 in PNAS, prompted protests around the world. In this study, for a period of a week, 689,000 Facebook users were sent news feeds that contained either an excess of positive terms, an excess of negative terms, or neither. Those in the first group subsequently used slightly more positive terms in their communications, while those in the second group used slightly more negative terms in their communications. This was said to show that people's 'emotional states' could be deliberately manipulated on a massive scale by a social media company, an idea that many people found disturbing. People were also upset that a large-scale experiment on emotion had been conducted without the explicit consent of any of the participants.

Facebook's consumer profiles are undoubtedly massive, but they pale in comparison with those maintained by Google, which is collecting information about people 24/7, using more than 60 different observation platforms – the search engine, of course, but also Google Wallet, Google Maps, Google Adwords, Google Analytics, Chrome, Google Docs, Android, YouTube, and on and on. Gmail users are generally oblivious to the fact that Google stores and analyses every email they write, even the drafts they never send – as well as all the incoming email they receive from both Gmail and non-Gmail users.

If Google set about to fix an election, it could identify just those voters who are undecided. Then it could send customized rankings favoring one candidate to just those people.

According to Google's privacy policy – to which one assents whenever one uses a Google product, even when one has not been informed that he or she is using a Google product – Google can share the information it collects about you with almost anyone, including government agencies. But never with you. Google's privacy is sacrosanct; yours is nonexistent.

Could Google and 'those we work with' (language from the privacy policy) use the information they are amassing about you for nefarious purposes – to manipulate or coerce, for example? Could inaccurate information in people's profiles (which people have no way to correct) limit their opportunities or ruin their reputations?

Certainly, if Google set about to fix an election, it could first dip into its massive database of personal information to identify just those voters who are undecided. Then it could, day after day, send customized rankings favoring one candidate to just those people. One advantage of this approach is that it would make Google's manipulation extremely difficult for investigators to detect.

Extreme forms of monitoring, whether by the KGB in the Soviet Union, the Stasi in East Germany, or Big Brother in 1984, are essential elements of all tyrannies, and technology is making both monitoring and the consolidation of surveillance data easier than ever. By 2020, China will have put in place the most ambitious government monitoring system ever created – a single database called the Social Credit System, in which multiple ratings and records for all of its 1.3 billion citizens are recorded for easy access by officials and bureaucrats. At a glance, they will know whether someone has plagiarized schoolwork, was tardy in paying bills, urinated in public, or blogged inappropriately online.

As Edward Snowden's revelations made clear, we are rapidly moving toward a world in which both governments and corporations – sometimes working together – are collecting massive amounts of data about every one of us every day, with few or no laws in place that restrict how those data can be used. When you combine the data collection with the desire to control or manipulate, the possibilities are endless, but perhaps the most frightening possibility is the one expressed in Boulding's assertion that an 'unseen dictatorship' was possible 'using the forms of democratic government'.

Since Robertson and I submitted our initial report on SEME to PNAS early in 2015, we have completed a sophisticated series of experiments that have greatly enhanced our understanding of this phenomenon, and other experiments will be completed in the coming months. We have a much better sense now of why SEME is so powerful and how, to some extent, it can be suppressed.

We have also learned something very disturbing – that search engines are influencing far more than what people buy and whom they vote for. We now have evidence suggesting that on virtually all issues where people are initially undecided, search rankings are impacting almost every decision that people make. They are having an impact on the opinions, beliefs, attitudes and behaviors of internet users worldwide – entirely without people's knowledge that this is occurring. This is happening with or without deliberate intervention by company officials; even so-called 'organic' search processes regularly generate search results that favor one point of view, and that in turn has the potential to tip the opinions of millions of people who are undecided on an issue. In one of our recent experiments, biased search results shifted people's opinions about the value of fracking by 33.9 per cent.

Perhaps even more disturbing is that the handful of people who do show awareness that they are viewing biased search rankings shift even further in the predicted direction; simply knowing that a list is biased doesn't necessarily protect you from SEME's power.

Remember what the search algorithm is doing: in response to your query, it is selecting a handful of web pages from among the billions that are available, and it is ordering those web pages using secret criteria. Seconds later, the decision you make or the opinion you form – about the best toothpaste to use, whether fracking is safe, where you should go on your next vacation, who would make the best president, or whether global warming is real – is determined by that short list you are shown, even though you have no idea how the list was generated.

The technology has made possible undetectable and untraceable manipulations of entire populations that are beyond the scope of existing regulations and laws.

Meanwhile, behind the scenes, a consolidation of search engines has been quietly taking place, so that more people are using the dominant search engine even when they think they are not. Because Google is the best search engine, and because crawling the rapidly expanding internet has become prohibitively expensive, more and more search engines are drawing their information from the leader rather than generating it themselves.

We are living in a world in which a handful of high-tech companies, sometimes working hand-in-hand with governments, are not only monitoring much of our activity, but are also invisibly controlling more and more of what we think, feel, do and say. The technology that now surrounds us is not just a harmless toy; it has also made possible undetectable and untraceable manipulations of entire populations – manipulations that have no precedent in human history and that are currently well beyond the scope of existing regulations and laws. The new hidden persuaders are bigger, bolder and badder than anything Vance Packard ever envisioned. If we choose to ignore this, we do so at our peril."

Which is precisely why we are doing this study and why we've entitled it, "Subliminal Seduction: How the Mass Media Mesmerizes the Minds of the Masses." As you can see, this is really happening on a grand global scale from a multitude of Media outlets and it's frankly sick and disturbing! You not only have devised a wicked scheme to "subliminally" control all information people are receiving for "brainwashing" purposes, but you are also "subliminally" determining people's so-called election "choices." Yeah, who's "choice" is it really?

In fact, as sad as those "manipulative" facts are, even secular researchers are catching up to the fact that this kind of "subliminal persuasion" and "silencing of dissenting views," and "controlling" and "monitoring" even "blacklisting" and "banning" of all information that appears online including on Social Media, is really going on 24 hours a day 7 days a week on a global basis, like this researcher shares.

Roya Ensafi, Assistant Professor Computer Science and Engineering University of Michigan: *"One of the core principals behind the design of the internet was that users from everywhere should get the same experience. Censored Planet is the first global internet censorship observatory that collects data continuously from more than 170 countries without end user's participation."*

Censored Planet uses advanced algorithms that can remotely detect network anomalies in order to track whether users around the world have access to thousands of important websites.

"The reason why I am interested to monitor and bring transparency and accountability to this phenomenon is that I strongly believe that it takes away opportunity, and it degrades human dignity for many who experience it. The technology behind internet censorship is becoming cheaper and cheaper, so it is very affordable for governments, states, you name it, to just buy this and deploy it.

There are three common ways that the governments block content: One is by blocking the IP address. There is also DNS poisoning where the DNS query is going to be poisoned or redirected to where the content is not there. The third kind is application layer blocking or keyword blocking where the censor monitors the traffic and notices there is sensitive keyword that is being communicated. As a result of that it's blocked or actually it shows the user a blocked page where it says this content is not legal. So, Censored Planet captures all of these three kinds of monitoring techniques."

Previous projects for monitoring global censorship have relied on volunteers around the world to collect data. But asking citizens in sensitive countries to participate could put them at risk. Censored Planet requires zero user participation, providing a safe method for tracking censorship everywhere so we can better understand what governments are blocking, and why.

"I grew up with the internet being censored either by university, by government, or more importantly by servers who provide the information. More and more I realized that internet censorship should be put into global perspective. It's not just Iran, just China. What I think is important is having a transparency on the process. Who is deciding to block what, and do we have trust in those people for the decisions they're making.[6]

And the resounding answer would be, "Absolutely not!" Based on their behavior, these people cannot be trusted, they're lying to us, they're deceiving us, manipulating us, even subliminally affecting our beliefs, buying habits, and even who to vote for! I didn't sign up for that, did you? But, as you can see, how much more proof do we need to "see" that these "Aliens" I mean rich "Elites" are literally "controlling" all forms of Media on the whole planet, including the worldwide web or Internet and all of Social Media, and we're only getting what these "Elite" owners "want" us to get, that is, their "preprogrammed narrative" that "subliminally" promotes their own nefarious agenda, not our desires, not the facts, not the truth!

And as bad as that is, it's about to get even worse! As horrible and rotten as what you just saw concerning the "manipulation" of Social Media by these "Aliens" I mean rich "Elites is, believe it or not, these same "manipulative" "Elites" are not only "controlling" all the Social Media content around the whole world, but they are actually "subliminally manipulating" us with a technology that's purposely "addictive" like Television, so we never "cut the feed" off from their daily "subliminal programming" of our minds! If you don't believe me, here's just some of that proof.

Adam Alter, author, Irresistible: *"Your phone is a dopamine pump. Behavioral addiction is a lot like substance addiction in a lot of ways, but it's much newer. So, substance addiction obviously involves the ingestion of a substance, and in the short-term that feels good, and in the long-term it harms your well-being in some respects. It can be physiological, it can be psychological, it can harm your social life, it can cause you to spend too much money, it can have a lot of negative effects on your life.*

Behavioral addiction is similar, the big difference though is that behavioral addiction does not involve ingestion of a substance, and it's much newer, it's a much more recent phenomenon. So, substance addiction has been around for a very long time, by some accounts for many thousands of years, but there weren't behaviors around that were compelling enough to rise to the level of addiction until quite recently.

The reason is that for them to be addictive, basically what has to happen is there's a behavior that you enjoy doing in the short-term that you do compulsively. So, you keep returning to it over and over again, but then in the long-term it harms your well-being. And it can, again, harm your well-being in lots of different respects, social, financial, physical, psychological. And I think the reason why we've got these new forms of addiction, there are two main reasons: The first one is that technology is much more sophisticated and advanced than it was even 20 years ago.

You're able to deliver the kinds of rewards that you need for a system to be addictive. So basically, what people are looking for is unpredictability and rapid feedback of either rewards (or if it's negative then negative experiences) and you actually need that mix of positive and negative feedback. Just as, for example, when you post something online, sometimes you're going to get a lot of hits, sometimes you aren't, and it's that unpredictability that we find so compelling.

You need to be able to deliver those rewards really rapidly, and for that you need the internet with the right kind of bandwidth to be able to deliver those rewards. The other thing that I think has happened is that companies are much savvier about this. There are employing behavioral experts to tell them how to design their media, how to design the vehicles that deliver those media, smart phones, iPads, smart watches, things like that. And for that reason, I think they are delivering products to us that are harder for us to resist. They've got enough features built in that we find to be pretty hard to resist, and then we end up developing addictions to them, and by some counts that applies to about half of the population of the developed world.

It turns out that it's not the addiction to the device itself. The device is just an incredibly effective way of delivering addictive content. So, when you're addicted to a screen, it's not that the screen itself is something that you can't get enough of. It's what it's providing. I think one of the reasons why we are so addicted to screens or to the content they provide is that they go with us wherever we go. And that's relatively new.

And so, if you played video games in the 80s or 90s or even the early 2000s, they didn't really go wherever you went as much as they do now, especially those games that were tied to the internet. Those were tied to your PC; you'd play where you were, and you didn't really leave with them as much. You had some portable devices, but those were much more primitive. Today with iPhones you can connect to other people on the go, you always have access to games, you always have access to email, you always have access to the internet, and you always have access to social media, and so they are great vehicles for providing the hits that you need when you need them.

Basically, we tend to develop addictions when we have a psychological need. And we get those whenever we're bored, whenever we're feeling a little bit lonely, whenever we're not really sure what to do with ourselves next, whenever we don't feel particularly efficacious (like we're having an effect on the world that we'd like to be having), those are the moments when you're looking for what some people call the 'adult pacifier.' And smart phones tend to be a great adult pacifier because after those moments you turn on your screen, you swipe, and you feel relaxed again. That's how people describe the experience.

So, they tend to be excellent devices for delivering these small hits that we look for, and social media is a great example. So, social media obviously now travels with us. It used to be confined to home computers to a large extent, but that's no longer the case. And people spend about three hours a day on average using their smart phones, which is pretty staggering. That is a huge chunk of the day of the waking hours that we spend when we're not at work. And what that means is they're spending a lot of the time returning over and over again to check Twitter, Instagram, Facebook, Snapchat and so on, and they're checking for a number of things.

One thing is that these media are bottomless, which means that you're constantly checking for new information; there's always something new to be kept up on. The other thing is if you happen to be a poster of content, you're very curious about getting rapid feedback as to whether people approve of that or they're not particularly interested in it. So, a lot of the time what we do is we return over and over again compulsively to see whether we're getting the positive feedback that we seek when we post content."

IQ Millennial Question: *"We know that engaging with social media and our cell phones releases a chemical called Dopamine. That's why when you get a text, it feels good. Right? We've all had it, when we feel a little bit down, or a little bit lonely, so you send out 10 texts to 10 friends, and get a high, high, high, high. Because it feels good when you get a response. That's why we count the likes. That's why we go back ten times to see if it's going, and if my Instagram is going slower, did I do something wrong, don't they like me anymore.*

The trauma for kids is to be unfriended. Because when we get it, we get a dose of Dopamine. It feels good, it's why we like it and why we keep going back. Dopamine is the exact same chemical that makes us feel good when we smoke, when we drink and when we gamble. In other words, it's highly, highly, addictive. Right? We have age restrictions on smoking, gambling and alcohol and no age restrictions on social media and cell phones. It's the equivalent of opening up the liquor cabinet and saying to our teenagers, hey by the way if this adolescent thing gets you down... But that basically is what is happening."[7]

Yeah, real funny! But he admitted, this is basically what's happening. Cell Phones, Social Media, Internet activities, are just as additive as alcohol and cigarettes and both of those are regulated and even have to have warning labels. But nothing for Social Media, the Internet, or Cell Phones? And we wonder why people can't seem to shut it off! It's simple! They got tricked into getting "hooked" on an addictive content, by "Aliens" I mean rich "Elites" who didn't tell anyone that they created this addiction "on purpose" called Social Media, the Internet and Cell Phones, just like they already did with Television. It's the exact same pattern! These "Elites" really are using these various new Global Media Technologies to keep people from ever turning away from their daily "brainwashing" procedures in order to not only get us to "buy" things to fill their pockets with our cash, to the tune of billions, but to "program" us and "reprogram" us into "believing" "behaving" "acting" and "thinking" whatever they want us to, including "who" we vote for! This is what Social Media is really all about, just like television. And now, globally, add to daily average home usage of television, nearly 7 hours, to the average time a person spends on social media, an additional 2 hours 24 minutes, with some people like teenagers, up to 9 hours a day, think about just how much the planet is being voluntarily "brainwashed" through these addictive platforms that they just can't seem to shut off on a global basis! And what are the results? The "Elites" have now created a society of global "brainwashed" addictive zombies, like this short film demonstrates.

The clip opens with the cartoon characters walking down the street, in a single file, staring at their cell phones, not looking where they are going and each one steps right into a manhole. A little child is sitting on the sidewalk looking at all the passersby's that all seem to be walking with their heads down looking at their phones. The child is amazed at what he is seeing. These people do not seem to have any idea of where they are going. On one corner a crowd has gathered to watch a fight going on. No one is trying to stop it, one guy is getting beat up by three other guys, but all the people are gathered around just to take pictures with their phones.

Then he passes by a restaurant and when he looks in the window, these people aren't even looking at what they are putting in their mouths. All they seem to be doing is looking at their cell phones. Even the baby in the highchair has his own phone to look at. As he goes farther down the street, he sees that the traffic has come to a stop. There is a fire in one of the buildings but what catches his eye is a lady taking a selfie with the fire blazing in the background.

He decides he needs to go home, and gets on the bus. But once again no one seems to know where they are going. All of them have their attention only on their cell phones. They are so involved in the phones that they don't even notice when a girl on the bus is getting bothered by a bully. No one was there to save her.

With the different apps that are found on the phones you can make yourself look like another person. He finally comes to a point he has to sit down and try to figure this all out. A little dog comes up to him and just as he is about to pet it someone not looking where they are going kicks the little dog farther down the street. That person had his head buried in his cell phone and wasn't watching where he was going or that he even kicked the dog.

It's like all these people have been mesmerized and are living in another world. People are taking pictures of their food with their cameras on their phones. He tries to get their attention, anyone's attention but no one even sees him. They are too busy reading, playing games, talking, taking pictures of the girl dancing in the street. They are like sheep going to slaughter or like people worshiping an idol. They all just follow the crowd and aren't really paying any attention to where they are being led.

A baby is born but crawls out of the room while the parents are taking pictures of themselves in the delivery room. They don't even notice that the baby is gone. A girl is on a ledge on a skyscraper. When she looks down, all the people are taking her picture on their phones. But no one seems to be concerned with her jumping. As she jumps their cameras follow her all the way to the ground, not missing a minute of the tragedy. The child sheds a tear. What has this world come to? It seems the whole population is following the crowd, with their eyes staring at their phones and they all walk right over the cliff.[8]

Wow! Looks like these "Aliens" I mean rich "Elites" have been successful in their takeover. They have now created, through their various forms of "addictive" Global Media, a society of "brainwashed" zombies who have no mind of their own, who will "do," "believe," "behave," "buy" and even "vote" for whoever these "Aliens" I mean rich "Elites" want them to. And lest you think these "Elites" don't know what they're doing to us, including the harm they're causing using their "addictive" technologies, this is precisely why these same Social Media giants refuse to let their own children have access to this same kind of Media. They don't want them "brainwashed" and messed up like the rest of us! Talk about hypocrisy! Check this out!

"The tech moguls who invented social media have banned their children from it. Silicon Valley parents are pulling the plug.

I have a huge TV in my living room, and I think we had a baseball game on in the background - we weren't even watching it or interacting with it. But then something curious happened; one dad, 'another figure in the tech industry,' was sitting on the couch, holding his baby, which began looking around the room, and its mother obscured its view of the screen, saying: 'No, you're not going to watch television at all until you're three years old.' She thought the baby being on this couch would be permanently damaging to her child.

This kind of behavior is becoming increasingly common as many of the tech world's leading lights, whose products have been used by millions of children the world over, are now intent on curbing their own offspring's screen time.

Not content with banning their children's devices, they are now legally stipulating that staff do the same. A report last weekend documented the rise in nanny contracts requiring that Silicon Valley sprogs not only be kept away from their own screens, but that those tasked with looking after them don't use their phones in front of the children, either.

Apple co-founder Steve Jobs was the first tech giant to admit, in 2011, that his own children had not used the recently released iPad created by his company, conceding that 'we limit how much technology our kids use at home.' And he wasn't alone: Microsoft founder Bill Gates set time limits on screens, banned mobile phones at the table and didn't let his children have them until they were 14, while Mark Zuckerberg implored his baby daughter to 'stop and smell the flowers' in an open letter to her which he released last year - one that made no mention of Facebook or even the internet.

Silicon Valley parents are strict when it comes to their children's screen time. Those at the very top, who were often barely out of university when they had their billion-dollar ideas, are now grown-ups with partners and children. Zuckerberg was still in his teens when he launched Facebook - now he's a 34-year-old father of two. Marissa Mayer, the former chief executive of Yahoo, was not yet 25 when she became employee number 20 at Google. She has since had three children. Jeff Bezos was 30 and newly married when he founded Amazon in his garage. He and his wife are now parents of four.

Yet it is unlikely that this will lead to a crisis of conscience, says Adam Alter, a professor of marketing at New York University and author of a recent book about technology addiction, because it would be 'completely inconsistent with the duty they have to their shareholders - to maximize profits.'

'For all the advantages they and their kids enjoy - from wealth to education - they don't trust themselves or their kids to be able to resist the charms of the very products they're promoting.' It would be 'silly' to expect them to change, he says. 'The best we can do is to try to uncover these hypocrisies and air them publicly.'"

Which is again, precisely why we're doing this study. Yet, as bad and hypocritical as that behavior by the "Elites" is, unfortunately, the negative and oftentimes dangerous side effects of Social Media is about to get even worse, at least for the rest of us. You see, this "subliminal" "zombie-like" "hypnotic" "addictive" form of Global Media called Social Media is not only being used to "control the narrative" of all information that we're receiving around the whole planet to "mesmerize the minds of the masses" for these "Aliens" I mean rich "Elites" and their nefarious purposes. It's also simultaneously creating the "immoral" society full of "destructive behaviors" that God warned about would come upon the scene when you're living in the Last Days! Let's take a look at what Social Media is now doing "morally" to people all over the planet! It's not good!

IQ Millennial Question: *"It's an entire generation that has access to an addictive, numbing chemical called dopamine through social media and cellphones as they're going through the high stress of adolescence. Why is this important? Almost every alcoholic discovered alcohol when they were teenagers. When we are very, very, young, the only approval we need is the approval of our parents and as we go through adolescence, we make this transition where we now need the approval of our peers. Very frustrating for our parents. Very important for us. That allows us to acculturate outside of our immediate families into the broader tribe.*

It's a highly, highly stressful and anxious period of our lives and we're supposed to learn to rely on our friends. Some people quite by accident discover alcohol and numbing effects of dopamine to help them cope with the stresses and anxieties of adolescence. Unfortunately, that becomes hardwired in their brains and for the rest of their lives, when they suffer significant stress, they will not turn to a person, they will turn to the bottle.

Social stress, financial stress, career stress, that's pretty much the primary reason why an alcoholic drinks. Right?

What's happening is, because we're allowing unfettered access to these dopamine producing devices and media, basically, it's becoming hardwired, and what we're seeing is, as they grow older, too many kids don't know how to form deep meaningful relationships. Their words, not mine. They will admit that many of their friendships are superficial. They will admit that they don't count on their friends. They don't rely on their friends, they have fun with their friends, but they also know that their friends will cancel out on them if something better comes along.

Deep meaningful relationships are not there because they never practice the skill set and worse, they don't have the coping mechanisms to deal with stress. So, when significant stress starts to show up in their lives, they're not turning to a person. They're turning to a device, they're turning to social media, they're turning to these things which offer temporary relief.

We know, the science is clear, we know that people who spend more time on Facebook have far higher rates of depression than people who spend less time on Facebook. If you are sitting at dinner with your friends and you're texting somebody who's not there, that's a problem. That's an addiction. If you're sitting in a meeting with people, you're supposed to be listening to and speaking and you put your phone on the table, face up or face down, I don't care, that sends a subconscious message to the room that you're just not that important to me right now.

That's what happens and the fact that you cannot put it away is because you are addicted. If you wake up and check your phone before you say good morning to your girlfriend, boyfriend or spouse, you have an addiction and like all addictions in time, it'll destroy relationships, it'll cost time and it will cost money and will make your life worse."

The Job Interview

Interviewer: *"Amy, it says you are trained in technology, that is very good.* (She is texting on her cell phone) *Are you adept at Excel?"*

Amy: *"No."*

Interviewer: *"PowerPoint?"*

Amy: *"No."*

Interviewer: *"Publisher?"*

Amy: *"Not really."*

Interviewer: *"Exactly in what area of technology are you proficient in?"*

Amy: *"Snapchat, Pinterest, Instagram, Vine, Twitter, you know the big ones."*

Interviewer: *"I'm surprised you didn't say Facebook."*

She laughs at that comment.

Amy: *"That's for old people, like my parents."* She laughs again. *"That's funny.*

Interviewer: *"Well, Amy when you are working for me you have to have those kind of research skills because I'll send you things for you to comb through and get the answers and send them to me. So, for that you've got to be really good at technology."*

Amy: *"For stuff like that, no problem, I'll just ask Siri."*

Interviewer: *"You'll just ask Siri?"*

Amy: *"You know, Siri tell me this; Siri find me that, we're all good at getting you the answers."*

Interviewer: *"Tell Siri I want you ready to go at 8:00 sharp each and every morning."*

Amy: *"I don't understand."*

Interviewer: *"What don't you understand?"*

Amy: *"What you just said."*

Interviewer: *"You don't understand, be ready to go?"*

Amy: *"No, you said 8:00, right?"*

Interviewer: *"Yes."*

Amy: *"8:00, like in the morning? 8:00?"*

Interviewer: *"Yes. In the morning."*

Amy: *"Yeah, that kind of doesn't work for me. Who gets up at 8:00?"*

Interviewer: *"I do."*

Amy: *"I Skype with my French boyfriend in Paris until like 3:00 in the morning. I don't even get to Starbucks until like 10:00 where I order my grande chai tea latte, three pumps skim milk, light water, 2% foam extra hot but not too hot. So, if it's okay I work best in the morning at 10:45."*

Interviewer: *"Wow, Amy, I don't think we're gonna be a good fit."*

Amy: *"Why are you being so negative? I can sense your hostilities and right now I am not feeling very safe. (She stands up to leave) I've been here for over five minutes and the only nice thing you have said to me was nice resume, which I typed all night for this meeting with you. You've given me no guidance, no validation, no encouragement, no supervision, is* there an HR Director somewhere?" (She is reading this from her cell phone)

Interviewer: *"HR Director?"*

Amy: *"Yes, I need to speak to someone, I may have to take off today as a mental health day."*

Interviewer: *"Take the day off? You, Amy, Amy look at me! You don't work here."*

Amy: *"Are you firing me?"*

Interviewer: *"Okay, yes."*

She takes a couple deep breaths and turns around and leaves.

Absolute Motivation: *"Imagine when you take that to extreme, where bad actors can now manipulate large swaths of people to do anything you want. It's a really, really bad state of affairs and we compound the problem, right? We curate our lives around this perceived sense of perfection because we get rewarded in these short-term signals. Hearts, Likes, Thumbs Up, and we conflate that with value, and we conflate it with truth instead of what it really is, it's fake brittle popularity that's short-term and that leaves you even more, and admit it, vacant and empty, before you did it because then it forces you into this vicious cycle where you're like what's the next thing I need to do now because I need it back. Think about that, compounded by two billion people and then think about how people react then to the perceptions of other. It's really bad."*

"So, we know from the research literature that the more you use social media the more likely you are to feel lonely or isolated. We know that the constant exposure to your friend's carefully curated positive portrayals of their life can lead you to feel inadequate and can increase rates of depression. And something I think we're going be hearing more about in the near future, is that there's a fundamental mismatch between the way

our brains are wired and this behavior of exposing yourself to stimuli with intermittent rewards throughout all of your waking hours.

So, it's one thing to spend a couple of hours at the slot machines in Las Vegas but if you bring the slot machine with you and you pull that handle all day long, from when you wake up to when you go to bed, we're not wired from it. It short-circuits the brain. We're starting to find that it has actual cognitive consequences, one of them being the sort of pervasive background hum of anxiety."

A Mother: *"I found an app. It was kind of like hidden. I went into it and I found photos on there of my daughter. She's 13 or 14, one more year is off the deep end."*

A Dad: *"I said where did you meet this person? He said, 'On Facebook.' He had a whole digital life that I hadn't even known existed."*

Female Student: *"I got to create my life the way I wanted to be seen and I would try to become that in real life even though it wasn't my true reality."*

Male Student: *"When I ever saw a friend request, I got so happy, I thought that everyone that I accepted was actually going to be a friend of mine."*

Doctor: *"50 percent of the kids when asked say they feel addicted to their social media."*

"We let the kids go into their own room doing their own things. We are busy, and we are tired, and we want to do our own thing."

Doctor: *"When you go into an average high school, one in four of those kids will admit to having sent a naked picture of themselves to a friend or stranger."*

Male Student #2: *"If your kid was smart, he could look up anything you want."*

Doctor: *"It's a free for all, there's pornography, there's constant reinforcement, pings, tings, new friends."*

Minister: *"We are concerned. We are really concerned about the influence of media, mass media, social media."*

Teacher: *"It's as new to us as it is to our kids. Switching from a teaching mode to a training mode."*

Male Student #2: *"They called me fatty. You're so dumb. Where they start getting you is where you believe."*

Doctor: *"That becomes their life. Social Media world but that world can be quickly abandoned."*

Male Student #2: *"I had a gun, at my head."*[9]

And what did God say? In the Last Days, your society would become filled with all kinds of absolute utterly dangerous destructive behaviors, to the point where they would not only be lovers of themselves, lovers of money, boastful, proud, abusive, disobedient to their parents, ungrateful, unholy, without love, unforgiving, slanderous, without self-control, not lovers of the good, treacherous, rash, conceited, lovers of pleasure rather than lovers of God, but even become "brutal." And that's the perfect word to describe what we're seeing today with people's behavior, thanks to "brainwashing" from all these emerging Global Media Technologies, including Social Media. It's the same rule and same pattern. Junk in equals junk out! If you ingest hours and hours on end of "immoral" and "addictive" and "dangerous" and even "destructive" content on Social Media, your life will begin to reflect it. And the next thing you know, you wake up with an extremely "immoral" society around the whole planet just like God warned about 2,000 years ago.

But unfortunately, that's only half of it! Remember the other consistent pattern here? These same "Aliens" I mean rich "Elites" are also using the Global Media of Social Media to create the biggest "scoffing" society in the history of mankind against God, Jesus, the Bible, and Christianity! See if you agree with just a few of the current headlines out there!

- Social Media Platforms Censoring Christians
- Is Facebook Censoring Christian Content?
- Social Media to Erode the Christian Belief System
- Facebook will go after conservative, Christian content.
- Christian Posts Blocked: Will 'Christian Speech' Be Allowed in the New Facebook World?
- Christian Organization REMOVED from Twitter
- Guy Banned from Twitch for His 'Religion'
- Christian Kevin Sorbo speaks out after Facebook removed his page.
- Twitter Ban: Censorship on "a way of talking" -- does that include Christian speech?
- Twitter Bans Bible Verses
- Did Facebook Really Ban the "I Am a Christian" Ad?
- Is Facebook Banning Bible Passages?
- Facebook bans, censorship and Christian faith
- Is censorship escalating against evangelicals?
- Facebook accused of 'censoring' Christian views.
- Christian TikTok Videos Are Censored and Deleted In The US
- Why is Twitter censoring Jews and Christians?
- Anti-Christian Censorship and New Media
- YouTube 'Cancels' Christian Media Network, Deletes Over 15,000 Videos
- YouTube ad policy bans keyword 'Christian'

Gee, I wonder why so many people around the planet are becoming so anti-Christian, anti-God, anti-Bible, and anti-Jesus these days? Could it be their "scoffing" attitude is encouraged by the anti-

Christian, anti-God, anti-Bible, and anti-Jesus "brainwashing" they're receiving from Social Media, whether they realize it or not? I think so! And by the way, that was just 20 of the "thousands" of recent headlines out there showing how "biased" and "anti-Christian" Social Media has really become! You "blacklist" them and "ban" them and "mock" and "scoff" at their Christian beliefs and their Biblical content and media. Again, whatever happened to "free speech?" Now, for those of you who might even be "scoffing" at this point thinking there's no way YouTube would ever "ban" the word "Christian," here's the proof! Check it out!

RT News Reports: *"A Veterans group in the US claims that YouTube has rejected its ad because it contained the word Christian, something the video sharing platform apparently deemed unacceptable. Chad Robichaux, the President and Founder of the group gives us his side of the story."*

Chad Robichaux: *"So, we've had different types of censorship in the past, maybe more suppression, but this was the first time that we had an ad denied for a specific keyword, Christian, which is a big part of our marketing. We are a faith-based organization that markets to reaching military guys who are struggling with post-traumatic stress, divorce issues, with their military service. We are not a political organization. We do not have a political agenda; we are a faith-based Christian ministry who are trying to reach a particular group of veterans who are struggling."*[10]

But not anymore! Google shut you down because you mentioned the word "Christian." Can you believe this? In the United States of America. Our "Christian" Founding Fathers must be rolling over in their graves after all their careful and deliberate attempts to promote Christian "beliefs" to "ensure" our "freedoms" which are based out of the Christian "Bible! I'll ask it one last time, whatever happened to "free speech?" But I guess if you "subliminally seduce" enough people to "scoff" and "mock" at the Bible, God, Jesus, and Christianity, then you can trick them into giving up their Biblical freedoms and go back to a tyrannical form of Government. It's almost like these "Aliens" I mean rich "Elites" knew what they were doing from the very beginning when they launched all these Global Media outlets we've been looking at! Go figure! But as you can "see" it's the same sick twisted pattern with all these different forms of Global Media, including Social Media. They're all being used to "control the narrative" and to "manipulate the minds" of the masses for the "Elites" "own" nefarious purposes and it's going on all over the world, whether you realize it or not! Total "manipulation" of the "minds of the masses," across the whole planet, 24 hours a day, 7 days a week, just by simply "owning" and "controlling" the Media called newspapers, radio, music, books, the education system, Television, and now even Social Media. Right now, as we speak, the whole planet is being bombarded by all seven and there's no escaping it! Or is there? That will be the discussion of our next and final section.

Chapter Eight

The Response to Subliminal Technology

By now, I'm sure you're feeling totally "beat up" like that previous movie scene with all this gut-wrenching information repeatedly punching you in the midsection via these "glasses" that this study is "forcing" you to "see" and deal with. But again, all this is for your good, and frankly, for the good of the planet. Acting like this Global Subliminal Seduction really isn't going on isn't going to help anyone, nor is it the proper response. Which brings us to the final issue. How do we respond to all this?

The facts are, we've been duped and "mesmerized" by all this Global Media Technology that these "Aliens" I mean rich "Elites" have been forcing us to use. In fact, the more we use it, the more it destroys our very lives and those around us.

"We convince ourselves technology is helping. The reality is people don't seem enough for us these days. We condense books into articles, into lists into gifs because who has time to read an actual sentence anymore? We scroll and click and double tap our lives away, wondering where the day

goes. We're more anxious than any generation before us, yet we keep downloading another app hoping that it could fix us.

We look to the latest trends, check the reviews on Yelp, investigate the hashtags, yet in a world with so much information we're somehow still missing the wisdom of life. We crave real connection, but we'd rather text about the hard conversations. We desire relationships but settle for following each other on Twitter. We stare at the pretty pictures on Instagram instead of the person standing right in front of us. We prefer filtered images to reality.

We listen to a new podcast over a person's day, we check social media more than we check in with our friends. In our constant need for new information, new insights, new everything, we lose the chance to make old memories. We like being busy. We wear it like a badge of honor. We can't focus because we're being pulled in 50 different directions. We can't commit because there always seems to be another option, a better alternative. We have trust issues with the people we've known for years, yet we trust someone's review that we just saw on Amazon.

We want to hold hands while still holding onto our phones. We scroll through Twitter while having lunch with our friends. We think we can keep one foot out the door while still falling for the person sitting in front of us. We always need the latest, the newest the most up to date, and then wonder why relationships don't last. We walk around stores on FaceTime and text during work because life without multi-tasking feels inefficient. We fill our schedule, we fill our days, we fill our phones, thinking that will cure the emptiness that we feel inside us.

We binge another series of that latest Netflix show because someone else's story drowns out the experience of our own. We complain that nothing feels real anymore but we don't give anything time to become real. We complain nothing has depth, but we don't allow things space to deepen. We're so quick to fill the silence or write someone off completely. We complain nothing feels authentic today as we select a new filter on our selfies. We recharge our devices but not ourselves, when we're out of time we just add more to our schedules.

We don't know who we want to become because we're too distracted by who everyone else is becoming. How many times has anyone ever said to you, I want your time? Whenever you hear that, know this truth. What people really want is your energy. Imagine if you gave someone an hour of your time but the whole time you were completely distracted on your phone. Versus, if you gave someone 10 minutes of your complete energy. I'm sure that any of us would select option 2, because what we're really searching for is energy. What we're really searching for is attention. What we're really looking for is presence. What we're really after is affection.

So, what we really need to do is give people our energy, not just our time. Take a moment. Take time to reflect. Take a step back. Who are the people that make you forget to look at your phone? Spend time with them. Starve your distractions, feed your focus. You can't do the big things if you're always distracted by the small ones.[1]

And those small ones are all the different forms of media that these "Aliens" I mean rich "Elites" have gotten us not only "glued" to but as we saw literally "addicted" to! It's ruining our very fabric of life, let alone "brainwashing" us to go along with their nefarious agenda! Which is to "buy," "sell," "obey," "never question," "reproduce," "sleep," and on and on it goes. Every day we are being distracted and "zombiefied" by their Global "mesmerizing" Media outlets called Newspapers, Radio, Music, Books, the Education System, Television, and Social Media, and it's all turning us into a planet of "isolated" individuals "willingly" eager to obey whatever our "Alien" I mean rich "Elite" overlords want us to obey! So, what's the response? Shut it off! Enough is enough! As the man said, "Re-engage with people." Become a "human" again! And that brings us to the proper "response" to all this "Subliminal Seduction" that's really going on all over the world, via the Global Media that these "Elites" are using to "mesmerize the minds of the masses."

First of all, can we shut it off? Is it too late? Are we helplessly "addicted" to this "brainwashing" media forever? No! There really is hope! You really can respond in an appropriate manner and be "free" from all this "mental manipulation" that's going on and that's precisely what we want to leave you with in this study. And that first response, that first step to "freedom" is simply this. You need to "admit" that we're living in a world of "illusion." The "Elites" have actually been telling us, flaunting it in our face for a long time, what they're doing to us, that is if you have the "glasses on" to "see." But the world of "illusion" they created for us is like this guy. Remember this movie?

Clip from the Truman Show

The newscaster is reporting that comets are still headed to earth. A couple of officers watching it say, "Wow, what else is on?" "Yeah, let's watch something else." The officers change the channel, and this is what comes on.

Announcer: *"Coming to you now from the largest studio ever constructed, it's The Truman Show." Jim Carrey, who plays Truman walks out his front door and waves at the neighbors and says, "Good Morning. Oh, and in case I don't see you, good afternoon, good evening, and good night."*

Christof, Creator of the Truman Show: *"What if no scripts, no cue cards.... What if you were watched every moment of your life?"*

Announcer: *"How many cameras do you have there in that town? I believe Truman was the first child to have been legally adopted by a corporation."*

Christof: *"That's correct."*

Interviewer: *"Brilliant."*

Announcer: *"What if everyone you knew was pretending."*

Actress: *"Hi Honey, look what I got at the checkout. Dishwasher safe."*

Truman: *"Amazing."*

Announcer: *"What if your world was make believe?"*

Christof: *"His world he inhabits is counterfeit."*

Christof's secretary: *"I'm not allowed to talk to you."* She whispers to Truman.

Truman: *"That's how I look on your tapes?"*

Christof: *"There's nothing fake about Truman himself."*

He didn't know it, until now. (A studio light calls through the glass dome covering Truman's little city) He looks up and sees the hole. He realizes that something is not right.

Truman: *"A lot of strange things have been happening here."* He tells his best friend.

Truman looks at his ring. Not knowing what he is looking at, it is a camera that has been put there to televise him in his daily routine.

Camera man: *"Do you think he knows?"*

Truman: *"I think I'm mixed up in something, something big."* He tells his friend.

Christof: *"We accept the reality with which we're presented."*

The secretary: *"Everybody's pretending to…"* But she couldn't finish her sentence to warn him.

His pretend wife: *"Truman!"*

As Truman realizes that things just aren't right, he starts to do a little investigating on his own and after seeing some big screen videos he says, **Truman:** *"The whole world revolves around me. Everybody seems to be in on it. I'm going to go away for a while."*

He realizes that his best friend has been in on this all along. While they are watching him on camera, his best friend tells him, "I'm not in on it, Truman, because the last thing I would ever do is lie to you." This is all being televised for the world to see. They fade the music.

Christof: *"That is our hero shot."*

But Truman is determined to leave. He gets in the car and proceeds to leave town. And he takes his pretend wife with him. They have everyone out searching for him. "Truman, where are you going?"

Somehow, he manages to get himself a boat, and the studio begins to cause a storm at sea with lightning and fog to try to scare him back to his home. While all this is happening, the crowds are cheering, hoping that Truman can make his escape.[2]

Good thing that was just a show, make-believe for our entertainment. Yeah, actually it's really what's being done to us. And we need to ask ourselves the same questions that the movie did. "What if you were watched every moment of your life?" "What if everyone you knew was pretending?" "What if your world was make believe?" "What if you didn't know it, until now?" You know, like after this study. And "Will you accept the reality with which you're presented?" Those are great questions, because the facts are, if you think about it, that movie, really has become our reality. With all the global "mesmerizing" Media that's been thrust upon us, every single day, we live in a make-believe world of "illusion" just like *The Truman Show*. Only, insert "your name" there! Make it personal, because that's what's going on! It's the Bob Show, or Susan Show, the Mike Show, the Carol Show, whatever, whoever your name is, put it in there, because that's what's going on, every day whether you realize it or not!

And so, the question is, "How do we respond to this? Or as *The Truman Show* puts it, "how will it end" for you? Well again, the first step is to simply "admit" that this is really going on. Don't run from it. Don't hide from it. Just deal with it. Open your eyes, keep them open and don't take those glasses off! That's the first and most important step. Don't go back to "sleep" and live in that world of "illusion" that the "Aliens" I mean rich "Elites" have created for us. You need to own up to it, and step two, the next response is to make the "commitment" to never go back again, like this movie premise reveals. Again, the "Elites" are telling us what they're doing to us if you have "eyes" to "see."

Clip from The Matrix

Morpheus: *"Do you want to know what it is? The Matrix is everywhere, it is all around us, even now in this very room. You can see it when you look out your very window. Or when you turn on your television. You can feel it when you go to work, when you go to church, when you pay your taxes. It is the world that has been pulled over your eyes to blind you from the truth."*

Neo: *"What truth?"*

Morpheus: *"That you are a slave. Like everyone else you were born into bondage. Born into a prison that you cannot smell or taste or touch. A prison for your mind. Unfortunately, no one can be told what the Matrix is. You have to see it for yourself. This is your last chance. After this there is no turning back. You take the blue pill, the story ends, you wake up in your bed and you be whatever you want to be. You take the red pill, and you stay in wonderland and I show you how deep the rabbit hole is."*

Neo reaches for the red pill and Morpheus speaks again. "Remember all I'm offering is the truth, nothing more. Neo reaches for the red pill, puts it in his mouth and swallows it.

The next clip is when the human form is breaking out of the pod, struggling to breathe for the first time. He realizes that he is connected to several tubes. He starts to disconnect the largest one from the back of his neck when he realizes he is not alone. He not only sees several at the same level as him but when he looks a little farther there are thousands of pods when humans inside that are still asleep.

As he is standing there in amazement, looking at all the other pods a large vehicle that looks like a bug flies up to him and grabs him. He can hardly breathe but the vehicle isn't there to harm him, it disconnects the large connection in the back of his neck and then turns around and flies back to where it came from. Suddenly all the cables that were attached to him start coming off him. He is free.

The bottom of the pod falls off and he falls through the hole through a long cylinder. It seems to never end when he comes to an opening and he flies through it and down into a large pond of water. He struggles to keep his head out of the water in order to breathe, and he sees an opening in the top and a large claw coming down to pick him up out of the water. It lifts him up through the opening in the ceiling and the doors close.[3]

As he rightly stated, "The Matrix is everywhere. Even now in this room." And so, the question is, how will you respond? "Which pill will you take?" Will you take the "blue one" after all you've seen in this study and go and put your head back into the sand? Act like this is not real and go back to "sleep" in this slave world of "illusion" that's been created for us, to become "mesmerized" and "brainwashed" and "power sources" for the "Aliens" I mean rich "Elites and their nefarious plans? Or will you "commit" to taking the "red pill" now that your eyes are "open" and "see" this Global Media Matrix for what it is and "commit" to be "lifted out of it" and "rescued" so you never go back again? That's your second choice or response with this study. Will you choose slavery or freedom? Which will you "commit" to?

Which now leads us to step three. How do we "respond" to this Media Matrix, this world of "illusion, this Truman like Show, that's been created for us by these "Aliens" I mean rich "Elites" around the whole planet, who are using this system for their own nefarious purposes? Well, it's simple. You not only need to "admit" and then "commit" to "freedom," but you need to now, step three, "take action." That is, you need to "shut it off." You need to "unplug" yourself from the system or Matrix. And you need to "right now" not tomorrow, not in the future, right now you need to "take action" "walk out" of the Matrix "shut the door" once and for all just like Truman did in his final scene. Let's take a look at that again.

Clip from The Truman Show

Truman has left the city and is now on a small boat out in the middle of the ocean. The water is calm, and he is enjoying being in the fresh air with the sun beating on his face. He is enjoying the smell of the ocean breeze when suddenly there is a large crashing sound and he ducks down into the body of the boat. He looks around to see what could have made this noise. The boat isn't moving, he walks to the front of the boat and he sees that the point of the boat has gotten lodged into a wall that is the same color as the water and the sky. He sticks out his hand to touch it.

As he looks around, he realizes that it is a solid wall with clouds and blue sky painted on it. It looked so real, but he is in a small pond surrounded by a wall. He hits the wall, but it doesn't budge. He collapses in defeat. Tears are falling as he finds himself in a fake world. He gets out of the boat and walks to a large flight of stairs coming up out of the pond. At the top of the stairs is a door that says 'Exit.' All the people that were watching him hurried out of the office and left the creator of the show there alone to try to smooth this situation over. As he is about to step through the door, a voice comes from overhead.

Christof: *"Truman, you can speak. I can hear you."*

Truman: *"Who are you?"*

Christof: *"I am the creator of the television show that brings hope and joy and inspiration to millions."*

Truman: *"Who am I?"*

Christof: *"You are the star."*

Truman: *"Was nothing real?"*

Christof: *"You are real. That's what made you so good to watch. Listen to me. There is no more truth out there than there is in the world I created for you. Same lies, the same deceit. But in my world, you have nothing to fear. I know you better than you know yourself."*

Truman: *"You never had a camera in my head!"*

Christof: *"You're afraid. That's why you can't leave. It's okay, Truman. I understand. I have been watching you your whole life. I was watching you when you were born. I was watching when you took your first step. I watched you on your first day of school. The episode when you lost your first tooth. You can't leave Truman. You belong here with me. Talk to me. Say something. Well, say something. You're on television. You're live to the whole world."*

All the public all over the world are sitting on the edge of their seats waiting to see what Truman does next. He turns around and faces the camera.

Truman: *"In case I don't see ya, good afternoon, good evening and goodnight."*

He turns around and walks through the exit door. The secretary that likes him, throws up her arms, puts on her coat and runs out the door to meet him in the real world. Cheers came from all around the world as they saw Truman escape.

The two patrol officers: *"He made it! Yeah! Go Truman!"*

Christof is stunned.

An employee: *"Cease transmission." And the program is off the air.*

Patrol officers: *"You want another slice? What else is on? Yeah, let's see what else is on. Where's the T.V. Guide?"*[4]

That's what you do. Step three, the proper response, you simply "take action" and you "cease transmission." You simply "walk out" of the Media Matrix that's been created all around you, like a cog in a wheel, for the mere "entertainment" of these "Aliens" I mean rich "Elites" and their nefarious purposes. Why? Because we're being "used" and "abused." That's why the proper response to regain our "freedom" is to first "admit" and then two "commit" and three "take action" or "walk out" and "shut" this Media Matrix "off," this daily "mesmerizing" this "brainwashing" and "programming" we've been subjected to through the Newspapers, Radio, Music, Books, Education System, Television, and even Social Media. Shut it off and don't go back, step three! And when you do, let me be the first to congratulate you and welcome you back to the "real world" a world of "freedom" "true freedom." It's good to have you here with us!

This is what I truly hope is your response to this "Subliminal Seduction" that's really "mesmerizing the minds of the masses" around the planet 24 hours a day 7 days a week, as you've seen throughout this study. But the greatest thing we desire for you is to not only choose "freedom" today, but "for all eternity" as well. You see, there's somebody else you need to understand and know about who's behind the scenes "directing" this whole Global Media Matrix system that's been "mesmerizing" the planet for nefarious purposes. It's not just "Aliens" I mean rich "Elites" it's actually a "spiritual" entity. And he's doing it not only for the mere "monetary" reasons like the "Elite" are but for "spiritual" reasons as well. That "other director" I'm talking about has not only really been "mesmerizing" you and "scripting" your every move, ever since you were a baby, but he's not human. He's Satan.

And the Bible is clear, he is the one who has been working with the "Elites" to "mesmerize the minds of the masses" and "blind their minds" to the truth of God's existence and the fact that we really can have a beautiful loving intimate relationship with the Creator of the Universe. Here is what the Bible tells us about his nefarious activities.

2 Corinthians 4:4 "Satan, who is the god of this world, has blinded the minds of those who don't believe. They are unable to see the glorious light of the Good News. They don't understand this message about the glory of Christ, who is the exact likeness of God."

    The Bible not only says that Satan is real, but that he's really out there "directing" this Media Matrix system around the whole planet that's "blinding" the "minds" of people keeping them in "bondage" from "seeing" the truth. He's doing it in collaboration, in these Last Days, with the "Aliens" I mean rich "Elites" via this Global Media Matrix system that been built to generate his "lies" that keep the whole world from "seeing" the truth about God's existence and the "good news" that we really can have a relationship with Him through Jesus Christ. This is why he's using the Global Media Matrix system in these Last Days to simultaneously "seduce you" into living an "immoral" life, against God and even "scoff" at the very idea of there being a God, so as to "blind" you from the "good news" that you really can be "free" from his grasp and have a relationship with God. It's all been carefully "scripted" by Satan using the latest Media Technology that God warned us about 2,000 years ago. Satan, along with these "Aliens" I mean rich "Elites" have built this system to "blind" and "mesmerize" the whole world from "seeing" God via their world of "illusion" that they've created and foisted upon us. Their "minds" are now "blind" to it.

    Which leads us to the final most important step you'll ever take in your life, and that is this, step four. You see, if you not only want "freedom" both now and forevermore, which I highly recommend, then you not only first need to "admit" then "commit" and then "take action" but you need to, step four, follow through and "receive" this "good news" you've been formerly "blinded" to from Satan. This is what he's trying to keep you from "seeing." That God is Holy, and we are not. And because of this, we're not headed to Heaven when we die, we're all disqualified, myself included! We've all blown it one way or another as the Bible admits.

Romans 3:23, 6:23 "For all have sinned and fall short of the glory of God. For the wages of sin is death, but the gift of God is eternal life in Christ Jesus our Lord."

And if you're anything like I used to be, you're probably thinking, "Well, I'm not that bad of a person, I'm not a sinner." Well, actually it's pretty easy to demonstrate. This is what God's Ten Commandments in the Bible are all about. They are God's X-ray showing us that we're really disqualified for heaven. For instance, how many of you, have ever lied, which is the 9th commandment, go ahead and raise your hand? Okay, for those who didn't raise your hand, you just did, you lied, because we've all done that at one time or another. Or how about this one, the eighth commandment, you shall not steal. How many of you have ever taken something that wasn't yours without permission, ever once, go ahead and raise your hand? Okay, you already told me you're a bunch of liars, so let's not commit another lie and not raise your hand. Because the truth is, we've all stolen or taken something without permission that didn't belong to us. That's just two out of the Ten Commandments. How are you doing? It's obvious, when you begin to see the X-Ray, that none of us can keep them. Which means we're all disqualified for Heaven. And the Bible says this about these kinds of people.

1 Corinthians 6:9-10 "Don't you realize that those who do wrong will not inherit the Kingdom of God? Don't fool yourselves. Those who indulge in sexual sin or who worship idols, or commit adultery or are male prostitutes, or practice homosexuality, or are thieves, or greedy people, or drunkards, or are abusive or cheat people – none of these will inherit the Kingdom of God."

In other words, go to heaven! We're not going to go there on our own! We're all disqualified! This is why Satan is using the Global Media Matrix system to "seduce you." This is the "seduction" into these kinds of "immoral" or sinful behaviors! He knows that it disqualifies us from

Heaven! But here's the "good news." The Bible says there's a way out of this dilemma. You can be "rescued" through Jesus Christ! That's why Satan has been "mesmerizing" the planet with this two-prong attack via the Media Matrix system he's built. One, he "seduces" us into a lifestyle of disqualifying "immoral" behavior. And two, he "blinds" our minds from "seeing" that God is not only real, but He's really provided a way out of this mess through Jesus Christ! This is what He says here!

1 Corinthians 6:11 "Some of you were once like that. But you were cleansed; you were made holy; you were made right with God by calling on the Name of the Lord Jesus Christ and by the Spirit of our God."

That's how we get out of this mess, this "bondage" from Satan and back into a life of "freedom" both now and for all eternity! This is step four. You "receive" this "good news" of being set free from Satan's bondage by "faith" and calling upon the Name of the Lord Jesus Christ and asking Him to forgive you of all your sins or "immorality" that Satan's tricked you into living. It's the only way out as Jesus says here.

John 14:6 "Jesus answered, I am the way and the truth and the life. No one comes to the Father except through me."

Romans 10:9-10 "For if you confess with your mouth that Jesus is Lord and believe in your heart that God raised him from the dead, you will be saved. For it is by believing in your heart that you are made right with God, and it is by confessing with your mouth that you are saved."

People, be encouraged, it really is true. If you would entrust your life to Jesus Christ and call upon the His Name and ask Him to forgive you of all your sins, then you too have become qualified for Heaven. I'm not going there because I'm perfect. It's simply because I've been forgiven! I "received" this "good news" by "faith" almost 30 years ago when my eyes were "opened" to "see" what Satan didn't want me to "see."

It's a "gift" from God, and that's why it's called the "gift" of eternal life. You can't earn it because we're all disqualified. Rather step four, you simply "receive" it by faith, what Jesus did for you, and you're on your way to Heaven once you do! Then the "blindness" drops off like this guy!

Acts 9:18 "Instantly something like scales fell from Saul's eyes, and he regained his sight. Then he got up and was baptized."

This is step four, the final and most important step you could ever take in life. This is how we hope you respond after reading this book. Please, "receive" Jesus Christ as your Lord and Savior today so the "blindness" created by Satan and the "Elites" will drop off like "scales" from your eyes. As the old hymn puts it, once you were "blind" but now you can "see." Then start reading the Bible because that's the only Book on the planet that not only came from God but has the power of God to "renew your mind" or in other words, "unbrainwash" you from the "brainwashing" you've been under via Satan's Global Media Matrix that's he's created in these Last Days. This is what the Bible says here.

Romans 12:2 "Do not conform to the pattern of this world but be transformed by the renewing of your mind. Then you will be able to test and approve what God's will is – His good, pleasing and perfect will."

The Bible really is the only Book on the planet that can "renew your mind" from the "brainwashing" and forced "blindness" you were under via the "Subliminal Seduction" that's been going on through the "Mass Media" that Satan and the "Elites" created to "mesmerizing the minds of the masses."

So again, I implore you, don't wait or put off till tomorrow to respond appropriately, because tomorrow may be too late. Please respond "today" with the truth you've encountered in this book and take "all" the steps to freedom. First "admit" then "commit" then step three "take action" and finally and most importantly, step four, "receive" the "good news" about having a relationship with God and going to Heaven to be with God through Jesus Christ. This is the "pill" that I extend to you one last time. Only it's not called the "red pill" it's the "Gos-pill" or "gospel," the "good news" of salvation through Jesus Christ. Once you "receive" it by "faith", it will be like you've awakened from the longest dream, as this man sings. I hope to see you there in Heaven. Respond today.

Words from the song "Your love Broke Through" by Keith Green

>Like a foolish dreamer trying to build a highway to the sky
>All my hopes would come tumbling down.
>And I never knew just why.
>Until today, when you pulled away the clouds.

>That hung like curtains on my eyes.
>Well, I've been blind all these wasted years.
>And I thought I was so wise.
>But then you took me by surprise

>Like waking up from the longest dream, how real it seemed
>Until your love broke through
>I've been lost in a fantasy, that blinded me.
>Until your love broke through

>All my life I've been searching for that crazy missing part.
>And with one touch, Lord, You just rolled away
>The stone that held my heart
>And now I see that the answer was as easy.

>As just asking you in
>And I am so sure I could never doubt.
>Your gentle touch again
>It's like the power of the wind.

>Like waking up from the longest dream, how real it seemed
>Until your love broke through
>I've been lost in a fantasy, that blinded me.

Until your love broke through
Like waking up from the longest dream, how real it seemed
Until your love broke through.[5]

How to Receive Jesus Christ:

1. Admit your need (I am a sinner).

2. Be willing to turn from your sins (repent).

3. Believe that Jesus Christ died for you on the Cross and rose from the grave.

4. Through prayer, invite Jesus Christ to come in and control your life through the Holy Spirit. (Receive Him as Lord and Savior.)

What to pray:

Dear Lord Jesus,

I know that I am a sinner and need Your forgiveness. I believe that You died for my sins. I want to turn from my sins. I now invite You to come into my heart and life. I want to trust and follow You as Lord and Savior.

In Jesus' name. Amen.

Notes

Chapter 1 — *The History of Subliminal Technology*

1. *History of Subliminal Technology*
   https://www.youtube.com/watch?v=DU-eUeXcaqk
2. *Effects of Subliminal Technology*
   https://www.youtube.com/watch?v=dS_kKbIL4dY
3. *Subliminal Logos*
   https://www.youtube.com/watch?v=dPpqXHWBy8Q
   https://www.youtube.com/watch?v=CHkkdDRqJOI
4. *Subliminal Advertising*
   https://www.youtube.com/watch?v=gDUCOj2NaMo
5. *Subliminal Cartoons*
   https://www.youtube.com/watch?v=iGHd3_xY1l0
   https://www.youtube.com/watch?v=YCd4zV735AM
6. *Subliminal Television & Movies*
   https://www.youtube.com/watch?v=aA-0BSzHXxw
   https://www.youtube.com/watch?v=J-PZMIATFYY
7. *Subliminal Technology Effects*
   https://www.youtube.com/watch?v=2lM9NH_hCqs
8. *Subliminal Technology Really Works*
   https://www.telegraph.co.uk/news/science/science-news/6232801/Subliminal-advertising-really-does-work-claim-scientists.html
9. *Definition of Neuromarketing*
   https://en.wikipedia.org/wiki/Neuromarketing
   https://www.waldenu.edu/programs/business/resource/how-neuromarketing-is-being-used-in-business-management
10. *Beaver Lies About Bike*
    https://www.imdb.com/title/tt0630196/
11. *Statistics of School Behavior*

http://www.lamblion.us/2012/04/decay-of-society-schools-as-mirror.html

Chapter 2   *The Methods of Subliminal Technology*

1. *Story of the Stranger*
   (Email Story: Source Unknown)
2. *Subliminal Smells*
   http://130.18.140.19/mmsoc/subliminal/
   http://www.mrcranky.com/movies/waterboy/71.html
   http://pf.fastcompany.com/magazine/97/brand-spirit.html
   http://www.cnn.com/2005/TECH/08/19/virtual.reality.reut/
3. *Subliminal Savors*
   http://www.ash.org.uk/index.php?navState=&getPage=http://www.ash.org.uk/html/./regulation/html/additives.html
   http://www.opencollege.info/health&nutritioncourse.html
   http://joi.ito.com/archives/2002/10/26/is_diet_coke_bad_for_you.html
   http://www.downtoearth.org/dtenews27/msg.htm
   http://www.vitalearth.org/nutrient_robbers.htm
   http://www.msgmyth.com/discus/messages/2/2.html?1123076295
4. *Subliminal Sounds*
   http://130.18.140.19/mmsoc/subliminal/
   http://www.mindcontrolforums.com/news/devious-things-with-your-mind.htm
   http://serendip.brynmawr.edu/bb/neuro/neuro04/web2/dyi.html
   https://serendipstudio.org/exchange/serendipupdate/are-you-being-brainwashed-muzak
5. *Subliminal Sights*
   http://130.18.140.19/mmsoc/subliminal/
   http://www.theage.com.au/news/tv--radio/doing-the-softsell/2005/08/23/1124562865635.html
   http://www.tsn.ca/auto_racing/news_story.asp?ID=131594&hubName=auto_racing
   http://www.commercialalert.org/issues-

article.php?article_id=429&subcategory_id=14&category=1
http://abc.go.com/primetime/xtremehome/featured/sears.html
http://home.earthlink.net/~patentlaw/rad.htm#prodmenu
http://theuncool.com/films/maguire/press/ewreebok.htm
http://www.hollywoodjesus.com/josie.htm
http://www.ccsg.it/pirelli.htm
http://www.ccsg.it/Dixan.htm
http://130.18.140.19/mmsoc/subliminal/candies.html
http://130.18.140.19/mmsoc/subliminal/denim.html
http://130.18.140.19/mmsoc/subliminal/jovan.html
http://130.18.140.19/mmsoc/subliminal/
http://www.ccsg.it/magliet.htm
http://www.ccsg.it/pepsi.htm
http://www.subliminal-message.info/
http://www.ccsg.it/silenzio.htm
http://www.ciadvertising.org/student_account/spring_01/adv391
k/hjy/adv382j/1st/application.html
http://www.subliminalsex.com/Subliminal%20TV%20Messages%20
Examples.htm)
http://www.ccsg.it/Kent.htm
http://www.sandrelli.net/subs3.htm
http://www.ccsg.it/ritz.htm
http://www.ccsg.it/bourbon.htm
http://www.ccsg.it/mentol.htm
http://www.poleshift.org/sublim/egs/S_e_x_Embeds.html
http://www.fpx.de/fp/Disney/Posters/little_mermaid_89_style_b.jpg
http://www.sandrelli.net/subliminals.htm
http://www.ccsg.it/walt4_S1.htm
http://www.ccsg.it/walt1_RR.htm
http://www.artistmike.com/Temp/SubliminalAd.html
http://www.ccsg.it/serpe.htm
http://www.asianjoke.com/cool_wacky_videos_about_asian.htm

6. *Subliminal Product Placement*
   https://www.youtube.com/watch?v=_3FLKuQa62c
7. *Subliminal Virtual Ads*
   https://www.youtube.com/watch?v=yG6nMgmeti4

8. *Subliminal Music Video*
   https://www.youtube.com/watch?v=PC9tgxm9BMM
9. *Subliminal Cartoons*
   https://www.imdb.com/title/tt0095776/
10. *Subliminal Political Ad*
    https://www.youtube.com/watch?v=rV2Fl50_e8A
11. *Subliminal Flashing*
    https://www.youtube.com/watch?v=QWv5jLO9g9Q
12. *Subliminal Interview*
    https://www.youtube.com/watch?v=NcvMgQvpRsc
13. *Subliminal Neuromarketing*
    https://www.youtube.com/watch?v=sVpSr5xJiNw
14. *Subliminal Psyche*
    http://www.atsnn.com/story/124672.html
    http://www.utne.com/webwatch/2005_188/news/11589-1.html
    http://www.commercialalert.org/issues-article.php?article_id=710&subcategory_id=82&category=1
    http://www.commercialalert.org/issues-article.php?article_id=709&subcategory_id=82&category=1
    http://www.commercialalert.org/issues-article.php?article_id=714&subcategory_id=82&category=1
    http://www.commercialalert.org/issues-article.php?article_id=707&subcategory_id=82&category=1
    http://www.commercialalert.org/issues-article.php?article_id=207&subcategory_id=82&category=1
    http://www.commercialalert.org/issues-article.php?article_id=708&subcategory_id=82&category=1
    http://www.commercialalert.org/issues-article.php?article_id=704&subcategory_id=82&category=1
    http://www.commercialalert.org/issues-article.php?article_id=712&subcategory_id=82&category=1
    http://www.commercialalert.org/issues-article.php?article_id=259&subcategory_id=82&category=1
    Patrick Frank, *Art Forms: An Introduction to the Visual Arts*, (Upper saddle River: Pearson Education Inc., 2004, Pgs. 60,61)
    David A. Lauer, *Design Basics*,

(Belmont, Thomson Learning Inc., 2005, Pgs. 120,226,246, 250, 262, 266)
Patrick Frank, *Art Forms: An Introduction to the Visual Arts*,
(Upper saddle River: Pearson Education Inc., 2004, Pgs. 60,61)
David A. Lauer, *Design Basics*,
(Belmont, Thomson Learning Inc., 2005, Pgs. 120,226,246, 250, 262, 266)
http://www.sondraslair.com/television.html
http://www.causeof.org/ianeurofeedback.htm
http://www.artlex.com/ArtLex/Su.html
http://faculty.washington.edu/~chudler/chvision.html
http://gummy-stuff.org/illusions.htm
http://www.artlex.com/ArtLex/a/images/afterimage_jesus.gif
http://www.ilusa.com/gallery2000.htm
https://www.businessinsider.com/10-biggest-advertising-spenders-in-the-us-2015-7#:~:text=Companies%20are%20constantly%20vying%20for,annual%20Leading%20National%20Advertisers%20report.

15. *Admission from Kids*
    Children, Youth and Family Consortium
    http://www.cyfc.umn.edu/Documents/C/D/CD1000.html

Chapter 3   *The Manipulation of Newspapers*

1. *Subliminal Technology Today Article*
   https://www.technologynetworks.com/tn/articles/how-subliminal-images-impact-your-brain-and-behavior-344858
2. *They Live Glasses*
   https://www.imdb.com/title/tt0096256/
3. *The Gutenberg Press*
   https://www.youtube.com/watch?v=OkfiT7p_1GA
4. *The New York times Printing Press*
   https://www.youtube.com/watch?v=MrWP2z8I0Qk
5. *History of Hearst*
   https://www.youtube.com/watch?v=HvJQ6AZfT6Q

6. *History of Yellow Journalism*
   https://www.youtube.com/watch?v=lUi5bsAhTk4
7. *Anti-Trump Bias Coverage*
   https://www.youtube.com/watch?v=SULEZ3Lvf2I
8. *Elitists Bias Use of News*
   https://www.youtube.com/watch?v=Z7ZmSxmhohs
9. *Billionaires Buy Up U.S. Newspapers*
   https://www.youtube.com/watch?v=lvDRV6Pichc
10. *Billionaires Buy up World Newspapers*
    https://www.youtube.com/watch?v=IbqhTKgv0hg
11. *Hearst Buying Other Media*
    https://www.youtube.com/watch?v=HvJQ6AZfT6Q
12. *Newspapers Information*
    https://www.livescience.com/43639-who-invented-the-printing-press.html
    https://www.localhistories.org/media.html
    https://en.wikipedia.org/wiki/Publick_Occurrences_Both_Forreign_and_Domestick
    https://www.worldometers.info/newspapers/
    https://www.localhistories.org/communicationstime.html
    https://timesofindia.indiatimes.com/when-and-where-was-the-first-newspaper-published/articleshow/2477418.cms#:~:text=Johann%20Carolus%20(1575%2D1634),as%20the%20world's%20first%20newspaper.
    https://www.4imn.com/top200/
    https://en.wikipedia.org/wiki/List_of_the_oldest_newspapers
13. *William Randolph Hearst Information*
    https://en.wikipedia.org/wiki/William_Randolph_Hearst#Criticism
    https://en.wikipedia.org/wiki/Yellow_journalism
14. *Political Media Bias*
    https://www.investors.com/politics/editorials/media-trump-hatred-coverage/
15. *Elitists Own Newspapers Globally*
    https://www.forbes.com/sites/katevinton/2016/06/01/these-15-billionaires-own-americas-news-media-companies/?sh=24045256660a

https://en.wikipedia.org/wiki/Concentration_of_media_ownership#Newspaper_and_Advertising
16. *Christian Media Bias*
    https://stream.org/anti-christian-media-bias-real-not-way-think/
    https://www.baylorpress.com/9781602584778/compromising-scholarship/
    https://nypost.com/2019/01/25/exposing-the-times-anti-christian-bias/
    https://www.freedommag.org/issue/201411-held-back/media-and-ethics/british-bias-corporation-religious-prejudice-media.html
    https://web.archive.org/web/20200928230249/
    https://www.winchestersun.com/2020/06/02/letter-media-biased-against-christians/

Chapter 4   *The Manipulation of Radio & Music*

1. *The History of Radio*
   https://www.youtube.com/watch?v=drLxfjqZHVo
   https://www.youtube.com/watch?v=qgkepUUED7k
2. *War of the World's Radio Broadcast*
   https://www.youtube.com/watch?v=Xs0K4ApWl4g
3. *War of the Worlds Broadcast Article*
   https://www.nydailynews.com/news/national/war-worlds-broadcast-caos-1938-article-1.2406951
   https://en.wikipedia.org/wiki/The_War_of_the_Worlds_(1938_radio_drama)#:~:text=Hundreds%20attacked%20Radio%20Quito%20and,of%20Paez's%20girlfriend%20and%20nephew.
   https://io9.gizmodo.com/real-life-casualties-from-war-of-the-worlds-373869
4. *Radio Preprogrammed 1975*
   https://www.youtube.com/watch?v=xBuTz3MJ2lQ
5. *AI Radio Disc Jockey*
   https://www.youtube.com/watch?v=wylpm7TxpIk
   https://www.youtube.com/watch?v=miSv2sIYlOc
   https://www.youtube.com/watch?v=za8TtziNfuw

6. *History of Satellite Radio*
   https://www.youtube.com/watch?v=3VPMeXFzWlw
7. *How to Make Subliminal Audio*
   https://www.youtube.com/watch?v=En9M2dYWN4Y
8. *Examples of Number Stations*
   https://www.youtube.com/watch?v=e6sE_kfNuKU
   https://www.youtube.com/watch?v=u0F984w4vLQ
9. *They Live Sleep*
   https://www.imdb.com/title/tt0096256/
10. *Backmasking Stairway to Heaven*
    https://www.youtube.com/watch?v=DgtxpRNT-r0
11. *Top 10 Music Subliminals*
    https://www.youtube.com/watch?v=AJLu8rkgi1I
12. *Musicians Admit Music Manipulation*
    (Video Source No Longer Available)
13. *Music Influences Behavior*
    (Video Source No Longer Available)
14. *Music Damaging Effects*
    (Video Source No Longer Available)
15. *Radio Information*
    https://www.learningwaves.ie/blog/243/in-an-era-of-fake-news-commercial-radio-goes-from-strength-to.html
    https://www1.udel.edu/nero/Radio/pdf_files/T&A_%20Media%20Ownership.pdf
    https://www.google.com/search?q=who+owns+all+the+radio+stations&rlz=1C1CHBF_enUS894US894&oq=who+owns+the+radio+stations+in+ame&aqs=chrome.3.69i57j0j0i39012.18556j1j15&sourceid=chrome&ie=UTF-8
    https://www.google.com/search?q=who+owns+the+radio+stations+in+america&rlz=1C1CHBF_enUS894US894&oq=who+owns+the+radio+stations&aqs=chrome.1.69i57j0j0i390.8588j1j7&sourceid=chrome&ie=UTF-8
    https://en.wikipedia.org/wiki/Media_cross-ownership_in_the_United_States
    https://en.wikipedia.org/wiki/Radio_homogenization
    http://www.insideradio.com/resources/who_owns_what/

16. *Subliminal Audio Messages Information*
    http://umich.edu/~onebook/pages/frames/usesF.html
    http://umich.edu/~onebook/pages/frames/historySet.html
    https://www.youtube.com/results?search_query=subliminal+self+help
    http://homepages.se.edu/cvonbergen/files/2012/11/Subliminal-Self-help-Messages_Do-They-Deliver.pdf
    https://www.mentalfloss.com/article/25989/10-facts-about-subliminal-messages-you-will-love
    https://en.wikipedia.org/wiki/Secret_broadcast
    https://www.vice.com/en/article/8x858k/the-mysterious-radio-stations-broadcasting-secret-messages
    https://coffeeordie.com/secret-radio-stations/
    https://en.wikipedia.org/wiki/Numbers_station
17. *Radio Anti-Christian*
    https://afajournal.org/past-issues/1994/september/tax-funded-anti-christian-bigotry-on-npr/
    https://afr.net/podcasts/the-hamilton-corner/2020/july/anti-christian-hostility-increases-on-account-of-the-word-of-god/
    https://ffrf.org/news/radio
    https://www.ewtn.com/catholicism/library/media-antichristian-propaganda-corps-9656
18. *Music Information*
    https://pop.inquirer.net/106559/the-auditory-phenomenon-called-backmasking-unmasked
    http://worldandi.misto.cz/_MAIL_/article/cijul99.html
    http://www.joesapt.net/superlink/shrg99-529/p1.html
    https://en.wikipedia.org/wiki/Radio_homogenization
    https://www.thefader.com/2017/09/18/music-labels-complicity-sony-umg-warner-trump#:~:text=When%20it%20comes%20down%20to,Group%2C%20and%20Warner%20Music%20Group.
    https://en.wikipedia.org/wiki/Media_cross-ownership_in_the_United_States
    https://en.wikipedia.org/wiki/Music_industry

Chapter 5   *The Manipulation of Books & Education*

1. *History of Books*
   https://www.youtube.com/watch?v=WrMJCHJt1Nw
2. *Magazines used to Manipulate*
   https://www.youtube.com/watch?v=peH6fBm0LD8
   https://www.youtube.com/watch?v=EU-wOf9Zs0k
3. *They Live Magazine Scene*
   https://www.imdb.com/title/tt0096256/
4. *Billionaire Bought Time Magazine*
   https://www.youtube.com/watch?v=RCbOqfCk6UY
5. *New Fahrenheit 451 Movie*
   https://www.youtube.com/watch?v=mNKwe9k55fs
   https://www.youtube.com/watch?v=nTz2-DyFRX4
6. *What is Political Correctness*
   https://www.youtube.com/watch?v=Om9bP3jNL7A
7. *Examples of Cancel Culture*
   https://www.youtube.com/watch?v=bisnMOujqFs
   https://www.youtube.com/watch?v=PDbi6XcnqSE
   https://www.youtube.com/watch?v=toh9R21GuQ4
   https://www.youtube.com/watch?v=lAftEcItyPo
   https://www.youtube.com/watch?v=FGCIW_wFx20
8. *Hitler Burning Books*
   https://www.youtube.com/watch?v=yHzM1gXaiVo
9. *Book Burning Today*
   https://www.youtube.com/watch?v=fwuCew0Dv6M
10. *Snowflake Generation*
    https://www.youtube.com/watch?v=C4apj9kDi2A
    https://www.youtube.com/watch?v=pP0WQTBKKBA
11. *Largest Church in America*
    http://www.youtube.com/watch?v=Qrqf2JasMJw
12. *Garfield & The Bible*
    http://www.youtube.com/watch?v=Qrqf2JasMJw
13. *Dumbing Down of America*
    http://www.youtube.com/watch?v=DDyDtYy2I0M

14. *The Truth About Inherit the Wind*
    https://www.youtube.com/watch?v=aaP7rj1_uSo
15. *No Intelligence Allowed*
    https://www.youtube.com/watch?v=vy5r-pDiZw8
16. *Kicked God Out*
    http://www.youtube.com/watch?v=mNjpddyn0HE
17. *Atheist Behavior Compilation*
    http://www.youtube.com/watch?v=r3cqPA0gJ-o
    http://www.youtube.com/watch?v=wAo_rEgR4xU
    http://www.youtube.com/watch?v=RYcOHrxNJjg&feature=related
    http://www.youtube.com/watch?v=KX7rhKfaG-A
    https://www.youtube.com/watch?v=uNn-9hw2YCc
    https://www.youtube.com/results?search_query=Jerry+Dewitt+Atheist+church+
18. *Book Print Media Information*
    https://www.youtube.com/watch?v=xpKqRC-9Avc
    https://askwonder.com/research/business-books-published-worldwide-year-annual-numbers-past-ten-years-ideal-qcrwjft2o#:~:text=This%20is%20based%20on%20data,approximately%202.2%20million%20per%20year.
    https://about.ebooks.com/ebook-industry-news-feed/
    https://www.markinblog.com/book-sales-statistics/
    https://www.businesswire.com/news/home/20200520005517/en/Outlook-on-the-Worldwide-E-Book-Market-to-2026---Key-Players-Include-Nokia-Amazon-Apple-Among-Others---ResearchAndMarkets.com
    https://www.theifod.com/how-many-new-books-are-published-each-year-and-other-related-books-facts/
    https://malwarwickonbooks.com/published-every-year/
    https://goodereader.com/blog/audiobooks/audiobook-trends-and-statistics-for-2020
    https://en.wikipedia.org/wiki/Media_cross-ownership_in_the_United_States
    http://publish.illinois.edu/englishadvising/big-five-publishers/#sthash.avfZXqMp.dpbs
    https://blog.reedsy.com/largest-book-publishers/

https://en.wikipedia.org/wiki/E-book
https://www.spyglassintel.com/visualization-of-circulation-revenue-for-the-top-12-us-consumer-magazine-publishers/
https://www.cnbc.com/2018/09/17/bezos-to-marc-benioff-why-billionaires-are-buying-media-companies.html
https://en.wikipedia.org/wiki/Fahrenheit_451

19. *Anti-Christian Book Banning*
https://kevincarson.com/2019/07/05/will-amazon-ban-christian-books-next/
https://www.overtoncountynews.com/lifestyles/california-bill-bans-the-bible/article_6477d400-6e74-11e8-9f20-87dc23ed0cd1.html
https://lawandreligionaustralia.blog/2015/05/11/schools-scripture-and-book-banning-in-nsw/
https://www.eternitynews.com.au/current/three-christian-books-banned-from-sre-curriculum-in-nsw/

20. *Education Information*
Vaughn Shatzer, History of American Education, Hagerstown: Word of Prophecy Ministries, 1999, Pgs. 3-9,12-13
http://www.wallbuilders.com/libissuesarticles.asp?id=8755
http://www.foxnews.com/us/2010/06/09/publishing-company-putting-warning-label-constitution/
http://www.SecularHumanism.org
http://www.lunarpages.com/stargazers/endworld/signs/occult.htm
http://www.linda.net/graphs.html
http://www.algonet.se/~tourtel/hovind_seminar/seminar_part1a.html
http://www.geocities.com/Heartland/Village/8759/youth-stats.html
http://www.biblesabbath.org/bacchiocchi/endtimewickedness.html
http://www.seebo.net/crisis.html
https://www.archives.gov/founding-docs/declaration-transcript
https://americanhumanist.org/what-is-humanism/manifesto1/
http://www.counterbalance.net/history/scopes-body.html
http://www.fillthevoid.org/Creation/Hovind/Brainwashed.html
http://cityroom.blogs.nytimes.com/2009/10/19/good-without-god-atheist-subway-ads-proclaim/
http://atheistbillboards.com/

http://www.theblaze.com/stories/florida-atheists-scrub-away-highway- blessing-with-unholy-water-because-theyre-not-going-to-tolerate-bigotry/
http://seattletimes.com/html/localnews/2015071658_Rapture17m.html
http://www.if-jesus-returns-kill-him-again.com/index.html
http://www.godblock.com/
https://www.goodreads.com/quotes/31896-i-am-afraid-that-the-schools-will-prove-the-very
http://tomohalloran.com/2014/07/15/hidden-weakness-pornogogues-running-childrens-schools/

Chapter 6   *The Manipulation of Television*

1. *History of Television*
   https://www.youtube.com/watch?v=PveVwQhNnq8
2. *Creating Dreams Like Netflix*
   https://www.youtube.com/watch?v=MP-hwl5d-gw
3. *The Matrix Artificial Womb Scene*
   https://www.youtube.com/watch?v=PKwq7b2i-vc
4. *Television Owned By Big Five*
   https://www.youtube.com/watch?v=A1_lCe3vyyc
5. *News Stations All Broadcasting Same*
   https://www.youtube.com/watch?v=_fHfgU8oMSo
6. *News Silencing dissenting Views*
   https://www.youtube.com/watch?v=p8gHVA4D7IQ
   https://www.cnn.com/videos/business/2021/02/03/newsmax-mike-lindell-2020-election-orig.cnn-business
7. *Television Look at Me*
   https://www.youtube.com/watch?v=e--nitVjLoQ
8. *Television Induces Hypnotic State*
   https://www.youtube.com/watch?v=D4eFdtxBjmw
9. *They Live Television*
   https://www.imdb.com/title/tt0096256/

10. *TV is Encouraging Immorality*
    https://www.youtube.com/watch?v=B7t85SESTXI
    https://www.youtube.com/watch?v=ijST6jWGMiM
    https://www.youtube.com/watch?v=p5sOOCb1JjM
11. *Ashley Madison Affairs*
    https://www.youtube.com/results?search_query=ashley+madison+commercials
12. *TV is Encouraging Scoffing*
    https://www.youtube.com/watch?v=p5sOOCb1JjM
13. *They Live Fight Scene*
    https://www.imdb.com/title/tt0096256/
14. *Television Media Information*
    https://www.quora.com/How-many-TV-channels-are-there-in-the-world
    https://kidshealth.org/en/parents/tv-affects-child.html
    https://en.wikipedia.org/wiki/Media_cross-ownership_in_the_United_States
    https://hightimes.com/culture/flashback-friday-subliminal-advertising/
    https://www.stanfordchildrens.org/en/topic/default?id=television-and-children-90-P02294
    https://www1.udel.edu/nero/Radio/pdf_files/T&A_%20Media%20Ownership.pdf
    https://www.webfx.com/blog/internet/the-6-companies-that-own-almost-all-media-infographic/
    https://www.ewtn.com/catholicism/library/media-antichristian-propaganda-corps-9656
    https://web.archive.org/web/19990502000502/http://www.cyfc.umn.edu/Documents/C/D/CD1001.html
    https://www.titlemax.com/discovery-center/lifestyle/who-owns-your-news-the-top-100-digital-news-outlets-and-their-ownership/
    https://www.amazon.com/Hollywood-Propaganda-Movies-Music-Culture-ebook/dp/B08KPQ1JNJ
    http://adage.com/century/rothenberg.html
    http://adage.com/news_and_features/special_reports/tv/1960s.html
    Ted Baehr, *The Media-Wise Family,*

(Colorado Springs: Chariot Victor Publishing, 1998, Pg. 19)
Ted Baehr, *The Media-Wise Family*,
(Colorado Springs: Chariot Victor Publishing, 1998, Pgs. 70, 71)
Ted Baehr, *The Media-Wise Family*,
(Colorado Springs: Chariot Victor Publishing, 1998, Pg. 78)
Ted Baehr, *The Media-Wise Family*,
(Colorado Springs: Chariot Victor Publishing, 1998, Pgs. 132-133)
Ted Baehr, *The Media-Wise Family*,
(Colorado Springs: Chariot Victor Publishing, 1998, Pg. 111)
Ted Baehr, *The Media-Wise Family*,
(Colorado Springs: Chariot Victor Publishing, 1998, Pg. 124)
Ted Baehr, *The Media-Wise Family*,
(Colorado Springs: Chariot Victor Publishing, 1998, Pg. 126)
Ted Baehr, *The Media-Wise Family*,
(Colorado Springs: Chariot Victor Publishing, 1998, Pg. 84)
Ted Baehr, *The Media-Wise Family*,
(Colorado Springs: Chariot Victor Publishing, 1998, Pg. 137)
https://www.focusonthefamily.com/lifechallenges/love-and-sex/purity/what-your-teens-need-to-know-about-sex
https://www.sagepub.com/sites/default/files/upm-binaries/23151_Chapter_6.pdf
http://screenrant.com/10-tv-shows-most-nudity/
https://en.wikipedia.org/wiki/Nudity_in_American_television
http://www.vulture.com/2014/08/cable-nudity-countdown-clock.html
https://en.wikipedia.org/wiki/Virgin_Territory_(TV_series)
http://www.mtv.com/shows/virgin-territory
http://deadline.com/2014/07/happyland-incest-mtv-craig-zadan-neil-meron-802510/
(CNN Broadcast – Married & Dating? Ad Campaign by Pro-Adultery Site – YouTube Video – Source Unknown)
(http://www.cyfc.umn.edu/Documents/H/K/HK1005.html)
(http://www.cyfc.umn.edu/Documents/C/D/CD1001.html)
(http://www.cyfc.umn.edu/Documents/C/B/CB1032.html)
Ted Baehr, The Media-Wise Family,
(Colorado Springs: Chariot Victor Publishing, 1998, Pg. 19, 70, 71)

Chapter 7   *The Manipulation of Social Media*

1. *The History of Social Media*
   https://www.youtube.com/watch?v=cw0jRD7mn1k
   https://www.youtube.com/watch?v=KRsix3bi0jU
2. *Facebook designed to Trace People*
   https://www.youtube.com/watch?v=iRT9On7qie8
3. *Facebook Tracing Compilation*
   Billy Crone, *The Final Countdown Tribulation Rising Vol.2 Modern Technology*, Las Vegas, 2019)
4. *Social Media User Totals*
   https://www.youtube.com/watch?v=Z1D7_NH5TBA
5. *Social Media Silencing Dissenting Views*
   https://www.youtube.com/watch?v=4lGGlbZRoV4
   https://www.youtube.com/watch?v=xIqYT3jB26c
   https://www.youtube.com/watch?v=rXtGItm1bvQ&t=2s
6. *Global Censorship Online*
   https://www.youtube.com/watch?v=ljZJrM9A2Pw
7. *Social Media Addictive*
   https://www.youtube.com/watch?v=NMq_MyOFtW8
   https://www.youtube.com/watch?v=sL8AsaEJDdo
8. *Cell Phone Zombies*
   https://www.youtube.com/watch?v=QugooaNRnsk
9. *Social Media Immoral & Destructive*
   https://www.youtube.com/watch?v=sL8AsaEJDdo
   https://www.youtube.com/watch?v=Uo0KjdDJr1c
   https://www.youtube.com/watch?v=PmEDAzqswh8
   https://www.youtube.com/watch?v=Hiltak3hsZc
   https://www.youtube.com/watch?v=dv2i7sYNiQ8
10. *YouTube Bans Word Christian*
    https://www.youtube.com/watch?v=gHIEdEIhgqo
11. *Social Media Information*
    https://www.page1.co.uk/blog/the-top-10-richest-social-media-entrepreneurs/
    https://www.statista.com/statistics/276312/net-worth-of-the-richest-

social-media-entrepreneurs/
https://backlinko.com/social-media-users
https://www.wvea.org/content/teens-spend-astounding-nine-hours-day-front-screens-researchers
https://blog.hootsuite.com/social-media-statistics-for-social-media-managers/
http://www.socialmediatoday.com/marketing/how-much-time-do-people-spend-social-media-infographic
https://drrobertepstein.com/index.php/internet-studies
https://www.usnews.com/opinion/articles/2016-06-22/google-is-the-worlds-biggest-censor-and-its-power-must-be-regulated
https://aeon.co/essays/how-the-internet-flips-elections-and-alters-our-thoughts
https://www.embopress.org/doi/full/10.15252/embr.202051420
https://www.independent.ie/life/family/parenting/the-tech-moguls-37494367.html
https://www.google.com/search?q=social+media+banning+Christian&rlz=1C1RXQR_enUS939US939&oq=social+media+banning+Christian&aqs=chrome..69i57.7128j0j7&sourceid=chrome&ie=UTF-8
https://www.youtube.com/results?search_query=socail+media+banning+Christian

Chapter 8   *The Response to Subliminal Technology*

1. *Global Media is Destroying Us*
   https://www.youtube.com/watch?v=ReyEU06e5PU
2. *The Truman Show Trailer*
   https://www.youtube.com/watch?v=dlnmQbPGuls
3. *Matrix Movie Red Pill Scene*
   https://www.youtube.com/watch?v=PKwq7b2i-vc
   https://www.youtube.com/watch?v=zE7PKRjrid4
4. *Truman Walks Out Scene*
   https://www.youtube.com/watch?v=Gn5kuDdeGzs
5. *Keith Green Your Love Broke Through*

https://www.youtube.com/watch?v=h89-3_kIRDA